PRACTICAL QUALITY MANAGEMENT
IN THE
CHEMICAL PROCESS INDUSTRY

INDUSTRIAL ENGINEERING

A Series of Reference Books and Textbooks

Editor

WILBUR MEIER, JR.
Dean, College of Engineering
The Pennsylvania State University
University Park, Pennsylvania

Additional Volumes in Preparation

PRACTICAL QUALITY MANAGEMENT IN THE CHEMICAL PROCESS INDUSTRY

Morton E. Bader

Chemicals Group
Olin Corporation
Stamford, Connecticut

MARCEL DEKKER, INC.   New York and Basel

Library of Congress Cataloging in Publication Data

Bader, Morton E.
    Practical quality management in the chemical process
industry.

    (Industrial engineering ; v. 7)
    Bibliography: p.
    Includes index.
    1. Chemical process control--Quality control.
I. Title. II. Series.
TP155.75.B334   1983   660.2'81   83-1844
ISBN 0-8247-1903-4

MARCEL DEKKER, INC.
270 Madison Avenue, New York, New York 10016

Current printing (last digit)
10 9 8 7 6 5 4 3 2 1

PRINTED IN THE UNITED STATES OF AMERICA

PREFACE

There have been an untold number of books and technical articles published over the years on the subjects of quality control, quality assurance and quality management.

To the author's knowledge, virtually all of these publications are concerned with the quality of hard goods. Many of these publications are heavy with statistical analysis. None, so far as can be determined, deals with quality programs for products manufactured by and for the chemical process industries.

Portions of this book were previously published in *Chemical Engineering*, as a guide to quality management for plant chemical engineers who had been given responsibility for plant product quality control, but who knew little or nothing about the subject.

This text has been enlarged to be helpful, not only to these chemical engineers, but to plant laboratory managers and those members of corporate or divisional staffs, who, having some responsibility for quality management programs, might wish to gain an overview of quality management.

The book is not intended to be an in-depth discussion of quality management philosophy for professionals with long years of experience in the field.

Some of the textual material is, indeed, fundamental. The author has made no assumption that the reader comes to the book with prior education or experience in the quality area.

The author has attempted to present technical information, both process and analytical, in a simple uncomplicated manner. Those statistical procedures discussed in the text are fundamental to laboratory use and are presented to be understood and useful even to those without extensive knowledge of statistical procedure.

The book is, in essence, a manual for designing and operating a basic quality management program. It is a practical discussion of what is needed and how to fulfill those needs on a practical basis.

Additionally, because packaging materials represent an important part of product quality, the text includes a discussion of currently available packaging components and how to monitor their qualtiy.

Because the chemical process industry is controlled by many federal regulations, and because product liability litigation is so much a part of industrial life and is associated with product quality, the author has included chapters discussing these topics.

The author is indebted to Edward J. Kuchar, Eugene S. Gazda and Chester A. Harlow for their valuable assistance and encouragement, and especially to Carol M. Didrich for her seemingly endless patience in the preparation of the manuscript.

Morton E. Bader

CONTENTS

INTRODUCTION

In very recent years product quality has become a topic for discussion in the press, on television, in the home and in industry. Quality of product is being treated as a new concept, recently discovered.

This has been occasioned no little by the startling success of Japanese industry, and to a lesser extent, by the West Germans. The quality of their manufactured goods is perceived to be high, and, as a consequence, much in demand, not only in the United States, but elsewhere in the world.

It is probably a truism that a company's product cannot endure the pressures of the market place if it is, or is perceived to be, of poor quality. As an extension of this, a company, regardless of its place in industry cannot remain viable if it manufactures products its customers cannot use or do not wish to accept for reasons of design, construction, utility, safety, purity or quality.

The first requisite to producing quality products is the total unwavering commitment of upper management to a quality pro-

gram. Without such commitment, any quality program will be sore-
ly tried and will never really succeed because of an innate weakness.

Consider, if you will, how your company might respond to
questions like these:

Do you have quality assurance and quality control designed
to produce material meeting specification, or, do you just
have a plant laboratory to which production personnel pay
no attention?

Do you have a staff quality function? To whom does it
report—upper management or two, three or more echelons
below upper management, where it can be safely tucked
away, out of sight.

To whom does the plant quality control laboratory report—
the plant manager, or a new young chemical engineer who
has little or no knowledge of, interest in, or commitment to
quality control. If the plant manager is too busy to bother
with quality control, maybe he has his priorities wrong.

Does management insist on having trained professionals in
its quality control programs, or does it accept semi-quali-
fied people because they may not be as expensive, or because they are members of the plant bargaining unit who
have opted, through seniority, to work in the plant quality
control laboratory?

What does your plant laboratory look like? Would you be
proud or ashamed to show it to your better customers?

Is management interested just in pounds, gallons or tons
produced, or in pounds, gallons or tons of product manufactured to specification— quality material?

The answers to these questions represent the attitude of
management to the production and sale of material which conforms
to its specifications and is fit for use by your customers. This attitude filters down through the organization and makes itself felt
by even the newest member of the company. When all else is said
what are you offering your customer but quality of product—and,
if not quality of product, what?

PRACTICAL QUALITY MANAGEMENT IN THE CHEMICAL PROCESS INDUSTRY

1
SPECIFICATIONS

I. INTRODUCTION

No company in the chemical process industries can stay in business very long if it ships products that its customers cannot use. Similarly, a company's life will be short if it purchases unsuitable raw materials for its processes.

This is singularly important in a marketplace which is almost constantly in flux and where basic raw materials are being substituted in processes by lower cost chemicals, sometimes without full knowledge of the effects of such substitution.

To ensure economic survival and foster economic growth, it should be a company's prime objective to produce products of quality designed to achieve maximum customer satisfaction at the most economical cost.

In order to achieve that portion of the twin goals of economic survival and economic growth which is derived from product quality, it becomes necessary to formulate appropriate policies

1

concerning quality and to implement these policies through various
procedures.

A commitment to quality must come from a company's upper
management. There is no substitute.

Lacking such a commitment, made without reservation by
management, it becomes increasingly difficult, and ultimately
impossible to produce and ship quality products on any consistent
basis.

It is necessary for each product, manufactured or sold, to
be identified by a specification. It is also necessary to provide
for the proper handling of the customer's order, the correct manufacture of the needed product and the orderly shipment of the
correct product in conformance with the customer's requests.

It is further necessary to have a procedure for determining
whether the customer is pleased with the product he receives;
and, if not, to provide a means for receiving customer feedback
and an orderly system for rectifying customer complaints.

To implement these policies and procedures requires an organization with clearly defined authority, responsibility, and
accountability, along with programs designed to achieve success.

Such programs involve both quality assurance and quality
control.

Quality assurance sets the policies, standards, methods and
specifications for monitoring the quality of production. Quality
control is the day-to-day (or hour-to-hour) monitoring of production for conformance to the standards and specifications set under
a quality assurance program.

Thus, quality assurance programs and quality control laboratories are the means by which a company keeps track of the properties of its supplies and products.

In recent years, quality programs have achieved increasing
prominence and importance. There are a number of reasons.

A sound quality program enhances the corporate image. Increasingly, institutional advertisements proclaim the merits
of companies' quality control programs.

A properly run quality program saves money. It is always
cheaper to do it properly the first time than to rework or
discard rejected production.

Increasing competition in the marketplace virtually precludes
the manufacture and marketing of inferior products. A well
designed quality assurance program can improve product

manufacture, while a well conceived quality control program can help keep off-specification products from being distributed.

Consumers, both corporate and individual, are becoming more sophisticated. They will no longer readily accept products of perceived or actual shoddy quality.

A variety of laws and regulations make it almost mandatory for a company to have at least a minimal quality control system.

Finally, quality control is something more than just internal process control. It is necessary to enable a company to meet its legal obligations — both regulatory and contractual. In many instances, it becomes the primary line of defense against costly litigation.

II.   SPECIFICATIONS

A.   General Specifications

The cornerstone of quality control is the specification. Specifications embody the control limits — the minimum and maximum values — of the various chemical and physical properties of both raw materials and manufactured products.
    While terminology may vary somewhat from company to company, there are generally:

Raw material specifications.

Manufacturing specifications (which may include pilot plant or R&D specifications).

Product (sales) specifications.

Customer specifications.

B.   Raw Material Specifications

Setting up these specifications is the essential first step in any system of quality control. The choice of raw materials, or

feedstocks, is determined in the early stages of research and production development.

At that time, the important properties of the raw materials
are determined and designed, and fairly flexible control limits are
set.

Raw materials — feedstocks — may be likened to chromosomes
and genes. What you put into the process determines what you
get out.

In the standard textbook illustrations of chemical reactions,
in which reagent grade chemicals are always assumed, the reaction products are clearcut, simple and straight forward. Balance
the equations and you're finished.

It doesn't happen quite like that in a chemical plant. The
raw materials are not always chemically pure; there are contaminants in varying concentrations from trace upwards.

The choice of raw materials will determine the purity of your
finished product, the by-product mix of your process, and waste
disposal problems, certainly a factor not to be treated lightly.

During pilot plant processing, the properties and control
limits are refined and made more definitive. It is at this stage
that the work of quality control begins. The properties of raw
materials critical to the manufacture of a product must be reviewed,
and methods of analysis either selected or evolved for these properties.

At this same stage in the selection of raw materials, it could
be a prudent decision to review the properties and process effects
of specific feedstocks which might be purchased from different
suppliers or which might emanate from different countries. Different processes used to produce a particular feedstock might well
vary one or another of the properties sufficient to have an effect
on your process or finished product, however slight.

The success of your process and the ultimate quality of your
finished product, in large measure then, depends upon the raw
materials used in the process. It will certainly be valuable to you
to take time at this early stage to determine the critical properties
of your raw materials, and their alternates, as well. Performing
this in a panic situation is ill-advised, costly and frequently un-
satisfactory for product quality.

Critical properties of raw materials will vary from material to
material and will be different for solids, liquids or gases.

A critical property can be anything that affects the useful-
ness of the raw material in the manufacture of other products.

This may include, for example, the presence of an unwanted
odor; a color variation that will impart an unwanted color to the

finished product; the presence of organic contaminants (such as $CHCl_3$ or CCl_4 in chlorine intended for water purification use); an incorrect particle size range of a diluent that could lead to segregation or stratification of particles in a dry mixture; or an incorrect pH that might cause undesirable precipitation during manufacture.

If, during the regular production process, say, the unwanted color or odor is removed by carbon-bed filtration, or the dry mix milled to provide particle size homogeneity, then these properties become less than critical and permit the raw material — or an alternate — to be used.

It is when the manufacturing process (or equipment available for the process) precludes use of raw materials with such problems that the properties become critical.

There must be analytical procedures, qualitative or quantitative, for all chemical and physical properties that need to be controlled in raw materials. The methods should measure not only desirable components but also possible contaminants or trace compounds.

The methods must be able to measure these properties to whatever accuracy is required by production needs (parts per million, or parts per billion, if necessary).

Not every raw material is used in a process to manufacture other materials. If we expand the concept of "raw materials" from process materials to any product purchased for use within the plant, the term can conceivably cover everything from paper towels to coal.

In a water purification plant, chlorine (which could be considered a raw process material in some plants) would be purchased as a finished product for purification use.

In our energy tight society, coal is a raw material that is becoming more important, more commonly used, and more expensive.

Specifications should be developed to include such properties as % moisture, % ash, % volatiles, % fixed carbon, % sulfur, and BTU/lb. The sampling procedures and analytical methods for coal can be found in the annual book of ASTM standards.

If these sampling procedures and methods for coal analysis are beyond the capability of your plant laboratory, it is wise to use one of the many commercial laboratories that routinely handle such analyses.

Do not forego setting up specifications for raw materials because the analytical methods are too sophisticated or expensive for your plant laboratory.

There are probably all too many small and medium size companies (and some large ones, too) in the chemical process industry
which pay little attention to the source or quality of feedstocks
used in their processes, preferring to rely upon data provided to
them by their suppliers.

This is not necessarily harmful in and of itself, but it is not
as satisfactory as determining raw material data, properties and
process usage for yourself; nor will general information provided
by a supplier tell you what the by-product mix will be in your
particular process, or how the by-product mix can change with a
substituted feedstock.

C.   Manufacturing Specifications

These specifications describe the chemical and physical properties
that govern the manufacture of a product. They describe the
product which your process will produce with your equipment,
raw materials and labor force.

Manufacturing (or production) specifications are used by
production workers as well as by plant quality control personnel
to assure that all manufactured products will meet the established
requirements of product quality. They should be designed to
establish finished product requirements, and represent minimum
quality performance standards for all plant manufacturing operations.

These specifications may originate in the R&D work on a
given product, and may be expanded following pilot plant production. At this stage, production managers will have identified
those properties that describe the product and, in some fashion,
distinguish it from others.

Also, it is at this stage that production and quality control
personnel should consult with each other concerning the final
manufacturing specification. Such consultation gives quality control a chance to discuss minimum and maximum tolerance levels,
analytical methods and instrumentation required to achieve them,
and the analytical workload necessary to control product quality.

If pilot plant information is insufficient to develop a manufacturing specification, or there exists a wide variation between
pilot and large scale production, it may be desirable to design a
"tentative manufacturing specification." Such a specification is
then subject to future refinement, but immediately serves to control the product in the early stages of manufacture.

Once production is in place and adequate control data have been assembled, the final specification can be designed. Bear in mind, at this point, that the tolerance levels set in the specification must be realistic. Setting minimums too low, or maximums too high, because of production or marketing pressures, may be self-defeating and lead to abnormally high production rejection by quality control.

Tolerance levels are those points above and below the desired norm at which production may still be accepted. Production outside this range must be rejected.

These levels are functions of the properties of the raw materials used in the process, the process itself, and the analytical capability of the control laboratory.

Consider an example of a raw material involvement:

If you are engaged in cutting back (diluting) a raw material to a finished product, the final concentration of the raw material is the critical number. If you are to dilute a 50.0% material to a 25.0% finished product, it is not enough simply to cut it in half. A 1% overage, to 25.25% would minimize the chances of producing product below 25.0%. However, a 1% under figure (24.75%) could also, for some products, be considered acceptable for minor quantities of production.

Specifications can usually be set up by using the guaranteed label claim as the minimum, and the label claim plus an overage of up to 1% as the maximum.

If the manufacturing process is tightly controlled and shows virtually no variances in day-to-day operation, a manufacturing specification overage of up to 0.5% can be considered as the maximum. For a process with wide variation, a plus tolerance of 1% (or even 2%) may have to be built into the specification.

In certain specific products, such as those covered by FIFRA (Federal Insecticide, Fungicide and Rodenticide Act) amended 1978, it is wise to consult the officials who police production of these products. This act, incidentally, covers not only pesticides but also chlorine and related products.

Another helpful organization is the Association of American Pesticide Control Officials which has proposed the following guidelines for judging approval of pesticide (including chlorinated compounds) formulations.

Pesticide active ingredient guarantee, %	Allowable deviation below guarantee
1.00	15% of guarantee
1.00 — 19.99	0.1 plus 5% of guarantee
20.00 — 49.99	0.5 plus 3% of guarantee
50.00 — 100.00	1.0 plus 2% of guarantee

Figure 1.

Also of importance in setting up specifications — raw material or manufacturing — is the analytical procedure to be used in checking them. If the specification claim for a property is to the second decimal place, the analytical method should be accurate to the second place as well.

If the accuracy and precision of the method are in the least doubtful, the maximum concentration (or overage) in the specification should be adjusted to reflect this uncertainty.

III.   DISPOSITION OF UNSATISFACTORY RAW MATERIALS

Ideally, raw materials not meeting specification limits should be rejected and returned to the vendor. Indeed the vendor's quality control procedures should have picked up off-specification material prior to shipment.

However, such action — returning materials to the vendor — could result in shutting down a production line or even temporarily closing a plant.

If the raw material is of such poor quality that its use would seriously and adversely affect the quality of the product, there is no choice but to reject it out-of-hand and risk the shutdown.

However, certain realities must be faced. Sometimes, a raw material may be only slightly out of specification — i.e., its variance may be in a property of secondary importance to the quality of the finished product. Such a raw material not only can be, but probably should be used, and the vendor advised of his shortcoming.

In such a case, the appropriate technical, purchasing and manufacturing people should be informed, and should be able to arrive at a decision regarding the use of the questionable raw material based upon quality control's advice.

IV.   ACCEPTANCE AND REJECTION OF PRODUCTION

When discussing acceptance or rejection of finished goods, another factor must be considered:

In addition to manufacturing specifications, there are, generally, another set of specifications issued by the sales or marketing departments. These are called, variously, sales or product or shipping specifications and represent what the customer has been told he is buying. These specifications are normally looser than manufacturing ones and allow for inherent variations in manufacturing processes. This, of course, permits a plant to ship product that is slightly out of manufacturing specification, but within product or sales specifications. Bear in mind that the only specification with which the customer is familiar is the sales specification.

V.   WAIVER SYSTEMS

From time to time, off-specification material will be produced in any plant, regardless of how well run it may be and how proficient and dedicated its personnel. This material cannot always be recycled or easily or economically discarded. However, it can, on occasions, be shipped to certain customers whose needs are not affected by the particular off-specification property.

In cases like this, the quality control laboratory will have rejected the material and informed both production and marketing personnel of this fact. Marketing (Sales) may then request a waiver of specification from a customer and, with his consent, approve shipment of the off-specification product to that customer.

Such a "waiver system" has advantages:

It precludes a customer complaint, since the customer is aware of the off-specification product he is buying.

It permits "honest" shipment of off-specification material.

It saves rework or dumping costs.

There is an opinion in quality control that too much time and effort is spent obtaining and using waivers, and not enough in

manufacturing product to specification.   The feeling is that if
there were no waiver system, the sheer cost of non-disposable
off-specification material would force manufacturers to produce
specification product.

    This argument is not without merit, but pertains more to the
"hardware" industry than to the chemical process industry. The
equipment used in "hardware" manufacturing is more precise, more
easily controlled and less subject to variation than is a chemical
process. The "hardware" manufacturers might well be able to do
without a waiver system, but the chemical process industry is not
yet at that point − and, indeed, may never be.

    When a product has completed its manufacturing cycle, it
may be relegated to a holding or quarantine area, awaiting approval by the laboratory. Some plants, lacking such an area, may
move finished product to an internal warehouse to await approval
and subsequent shipping.

    This is perfectly normal, but what happens when and if some
of this product is not approved by the laboratory? How is it to be
handled?

    If practicable, a separate area for storage of off-specification
product should be set aside and clearly marked as such. In addition, each container, carton, drum or pallet (whichever is most
convenient) should be carefully marked as rejected material that is
not to be shipped.

    Finished product that is approved for shipment by the quality
control laboratory should be in a secured container, correctly
labeled, accurately weighed and clearly marked as to batch or lot
number.

VI. PRODUCT IDENTIFICATION

Every product shipped from a plant should certainly have an
approved label that names the product in the container, and contains a statement of contents and concentration of the major components, as well as cautionary statements, including those required by law, regarding product handling and use.

    There must also be, on the label, an EPA Registration number
for the product (if required) and an Establishment number.

    Because of increasingly stringent government controls on
materials offered for sale, and because of increasing governmental

and consumer insistence that materials or products not deemed safe
be recalled from the marketplace, it becomes important and, indeed, necessary to further identify the product by date of manufacture, packaging or shipment.

This identification, in any of the forms noted above, must
appear on every container shipped from the plant. This includes
drums (both metal and fiber), corrugated shipping containers,
and the consumer-sized individual packages within the shipping
containers.

The code should include, in any acceptable variation, the
year, month and day of manufacture, packaging or shipping.
Where desirable a plant code-letter designation, a shift code, or
other date may be added. There are a number of automatic coding
devices available to perform this task easily and inexpensively.
The packaging and shipping date code designations are to be used
on materials representing multibatch or several shifts or days of
manufacture stored commonly in one silo or tank, from which is
drawn the product to be shipped.

In the event a product recall becomes necessary or desirable,
only those containers that are immediately involved need to be
isolated, destroyed or returned to the plant.

Bulk shipments can usually be identified by tank car, tank
truck or hopper-car number and date of shipment. This form of
shipment obviously does not lend itself easily to recall, but identification could lead to isolation of the product until a determination can be made regarding its disposition.

Considering the deteriorating status of rail shipments and
not uncommon truck accidents, identification of the product contained in the shipments may well minimize public safety and health
problems.

Legibility of code numbers on containers is frequently a
problem which is just as frequently ignored by plant personnel.
It serves no purpose to apply a date code to a container and have
only a portion of it sufficiently legible to read.

When this occurs, quality control personnel should notify the
supervisor in charge of that function. The stamp pad may lack
ink, the stamp may be worn, the roller wheel of an automatic system may be worn or dirty, stencils may be worn or torn, etc.

Whatever the cause, the problem must be corrected quickly.
Laboratory personnel should make periodic and frequent audits of
preshipment storage areas or the plant warehouse to check for
both the presence and legibility of code markings on containers.

VII.   PRODUCT QUALITY COMPLAINTS

Product quality complaints are virtually inevitable, even in the "best" plants with "good" product quality. Sometimes, customer complaints increase significantly during a buyer's market and, similarly, decrease during a seller's market. This is relatively common and should not cause undue alarm.

You will likely find, too, that complaints involving handling, packaging and shipping may outnumber product quality complaints by as much as two or three to one. It would appear that good process control is easier to attain than the control of the handling, packaging and shipping functions.

Invariably, a product quality complaint will end up in the quality control laboratory for verification of the claim. The complaint should be accompanied by:

A description of the complaint.

A sample of the product in question.

The code number of the product.

The order number of the shipment.

The date of shipment.

The mode of shipment (e.g., rail, truck, barge).

The laboratory, for its part, should seek out the retained production or shipping sample of the material in question, as well as the quality control data assembled at the time the product was manufactured and tested.

Both the complaint sample and the retained sample should be reanalyzed for the properties under complaint, and the results forwarded to those people in your company who respond to complaints.

Sometimes, it happens that the customer is not using the same analytical method as you, or has varied, in some way, the analytical method which you supplied him for your product. His laboratory people may not be as adept in using the analytical method as you. Any, or all, of these issues can be a causative factor in the rejection of your product and the cause of a customer complaint.

Sometimes, faulty or dirty sampling equipment used to sample a shipment can be the unwitting cause of a customer complaint; as can a tank car not thoroughly cleaned from a previous shipment.

Don't be reluctant to discuss these problems with your customer. He doesn't enjoy complaints either.

There are other types of complaints that, by their nature, should not end up in the control laboratory. These deal with shipment of wrong materials, billing errors, shortages, incorrect shipping modes, late deliveries, damaged containers, etc.

2
THE QUALITY CONTROL LABORATORY

I. PLANT QUALITY CONTROL LABORATORY

The basic department in implementing a quality assurance
program is the plant quality control laboratory. This may be anything from a small, one analyst affair with modest equipment, to a
large multi-analyst laboratory having much sophisticated instrumentation.

Whatever the physical size of the laboratory, there must be
sufficient instrumentation and equipment to do the job required.
Anything less defeats the laboratory's whole purpose.

Whatever the purpose of the laboratory, the staff, small or
large, must be competent and trained to fulfill that purpose. The
use of semi-skilled technicians or analysts to operate relatively
sophisticated analytical instrumentation will not always produce
dependable data.

At a small plant site, there will generally be a principal
quality control laboratory, sized and staffed, for that plant's
needs. There are, however, many plants, usually large ones,

15

that because of their size, complexity, or distance from the principal quality control laboratory, need to employ satellite laboratories to control production quality on a frequent and continuing basis.

These satellite facilities may sometimes be referred to as "process control laboratories."

They are generally small (one or two person) facilities located close to the process they are to control. And they are usually furnished only with that equipment and instrumentation essential to their purpose.

A word of caution about these laboratories. They have a common problem caused by their location. Because they are usually in, or adjacent to, the production area they serve, they may easily become contaminated with the very product they are to monitor, especially in the area of trace components.

Particular care must be exercised to keep glassware, equipment and instrumentation clean; otherwise, it is likely that analyzed results will become distorted.

More often than not, process quality control laboratories operate under stringent time constraints. There are frequent samplings routinely made during the production cycle, as well as those done during production upsets, and samples requested by foremen, supervisors or operators to satisfy their need to know.

It is wise to staff these small laboratories with skilled, patient technicians or analysts able to perform repetitive procedures quickly, often and accurately.

II. LABORATORY EQUIPMENT

Basic laboratory equipment for everyday use should include an adequate supply of assorted glassware, reagent chemicals, analytical balances, hot plates, water bath, fume hood, digital pH meter, refrigerator, titration equipment and a drying oven.

Specialized equipment is a function of the analytical workload of the laboratory. Such equipment includes gas and liquid chromatographs, infrared spectrometers, color meters, atomic absorption spectrometers, automatic titrators, and automated chemical analyzers.

Chromatography is probably the most widely used analytical technique. There is a fundamental difference between gas and liquid chromatography. Simply stated, the mobile phase in liquid chromatography is not an inert gas, but a liquid carrier which can be used to improve the separation ability of the system.

An additional piece of specialized equipment gaining in use
and favor is an integrator attached to the recorder of a chromato-
graphic system. It is used to measure the area under the signal/
time curve. There are plotting integrators, digital integrators
and, more recently, highly sophisticated digital integrators with
computer capability. The latter can operate several instruments,
switch instruments at preselected times, perform complex analyses
and manipulate statistical data.

These units are not inexpensive, but even in a modest sized
laboratory can be cost effective.

You should be aware that many, if not most, plant laboratories are not as described or pictured in laboratory furniture
catalogs. Indeed, it is not absolutely necessary for them to be of
picture book quality. There must be, however, sufficient area -
aisle width, bench space, etc. for each analyst to function effi-
ciently and safely. The laboratory itself must be well ventilated −
in addition to fume hood ventilation. This is an absolute essential.
No compromise should be permitted.

Space should be set aside for semi-permanent, or permanent
glassware set-ups, so that these do not impinge on daily work
space, yet are convenient to use.

Reagents, glassware, and standard solutions should be kept
in convenient locations to save time and eliminate waste motion.

III.   ANALYTICAL METHODS

There are a number of good sources of analytical methods commonly used in laboratories. If you already have a control laboratory,
you, obviously, are using certain methods.

These methods − any analytical methods to be used − should
be in printed, bound form, readily available and centrally located
for reference by the analysts, and permanently protected from
staining or chemical disintegration with plastic covers for each
page. Single copoes of these methods may be used at the work
bench.

There are a number of good sources of analytical methods for
use in control laboratories. I have listed a number of these at the
end of the chapter. Admittedly, this is far from a comprehensive
listing. However the EPA and NIOSH manuals should be a part of
every laboratory. Not only are they invaluable for the areas they
cover, but they contain the only government-approved methods
for these areas. They represent up-to-date methodology and may

be far more modern than the methods you may have inherited from the chemists who preceded you. Consider these sources. Among them you may find methods that are more accurate and precise and yield better reproducibility than those you now use.

They may be faster or, in the case of instrumental analysis, permit more samples to be run per unit time.

In a later chapter (5), there will be a discussion on how, statistically, you may determine the precision between two different analytical methods; perhaps, the one you have been using and a newer or different one — or for that matter, a modification of an old method.

IV. LABORATORY NOTEBOOKS

The data obtained from these analytical methods not only should, but must be, entered in ink in bound laboratory notebooks. The notebooks, preferably, should have sequentially numbered pages.

All data, including solution standardization, normality, sample weight, etc., and calculations, must be indicated. In these days of the ubiquitous electronic pocket calculator, the arithmetic used to calculate data is erased in the machine and not shown in the notebook. This is poor policy and should not be permitted. It is all too easy to tap the wrong number button, or transpose a number, or misplace a decimal when using a calculator, while the error may go unnoticed.

If the calculations are noted in a book, they are more susceptible to checking. Errors may be found more easily.

It should be absolutely forbidden, as laboratory policy, to enter data on loose scraps of paper — whether in ink, pencil or anything else. It should also be standard policy to date each day's results and have the analyst sign at the bottom of the page.

It is preferable, in a multi-analyst laboratory, for each analyst to have his or her own data book and be held completely responsible for the information contained in it.

Obviously, it is not enough to keep the data in a book. This quality control information was developed to be transmitted to production or shipping people for their use.

If quality control reports are issued by the laboratory as a means of communicating production quality or shipment quality data to interested parties, they should be made, if practicable, on preprinted and numbered forms.

Such forms should contain adequate space to show physical and chemical data, space for pertinent quality comments where appropriate, and a place for the date and the analyst's signature. Of course, these reports should be prepared — at the very least — in duplicate.

The laboratory responsibility for these data does not end with entry into a book or onto a report sheet. The data, all of them, must be retained and stored for easy retrieval. The period of retention is subject to discussion.

V.   ANALYTICAL DATA RETENTION

Some data, according to various federal regulations, must be retained for as long as twenty years; some for as little as two years. It is sound practice at this time to retain all finished product data indefinitely. Even in a large multi-analyst laboratory, data retention requires a relatively small area; and in today's political, social and economic climate, one never knows when quality control data will be needed to defend or explain a position.

VI.   SAMPLE RETENTION

Production or shipment sample retention poses a variety of problems. Some sample retention is mandated by federal regulation, some by good business practices. Certainly a sample should be retained for the useful life of the product, including storage prior to shipment from the plant, distribution storage, time in the market pipeline, and the consumer's storage and use period.

Obviously, samples of some production — e.g, gases, or product that deteriorates in standing — cannot easily be stored for long periods. Additionally, storage of even 4-8 oz. samples, in suitable containers, can take up substantial space over a period of, say, two years.

The choice of container size, type and construction is yours and depends largely upon your product.

Polyethylene, PVC, glass, metal, kraft, fibre, etc., and variations of these materials would all be suitable if they can contain the product safely over the required time period.

Sample containers should be capable of being clearly and permanently labeled with all information pertinent to the sample. Ideally, the samples should be stored on shelves in a separate building — ventilated and illuminated, with provision for water and

drainage, or legal disposal, should breakage occur and flushing
be necessary. The latter is important; don't overlook it.

Again, good business practice dictates a policy of sample retention. How much and for how long needs to be determined at
each plant. Do not be short-sighted and minimize the program in
the interests of near-term economy. A retained sample has molli-
fied more than one consumer and salvaged more than one possible
quality problem.

VII.   START-UP PROBLEMS

The start-up of a new process — or a significant change in an old
process — in a plant often places a considerable strain upon the
resources and facilities of the plant's quality control laboratory.
Much of the standard routine must be set aside; often, personnel
from other plant laboratories within the company are pressed into
temporary duty.

In any event, whether or not additional personnel are available, a great deal of extra analytical work will have to be done,
rapidly, on a vast number of process samples.

This should be prepared for well in advance of the start-up,
or process change, date. If new analytical methodology is involved, one should order and prepare those reagents that will be
used in the analytical procedure. The method, itself, should be
run repeatedly on special samples until ease and familiarity and
reproducibility of results is achieved.

Purchase any special glassware, and practice new laboratory
set-ups. If new analytical instrumentation is involved, purchase
it well in advance. Remember there may well be a several month
lead time in acquiring new instrumentation. When the instrument
arrives, check it out thoroughly, and begin to train analysts in
its use at the earliest opportunity. The instrument supplier can,
and frequently will, help out at this point, by assisting in the
training procedure.

With "spiked" laboratory-prepared samples, the accuracy and
precision of the new method can be determined. Production personnel should be made aware of this training program and its
progress, if for no other reason than to prevent questions of
credibility when early (and sometimes unfavorable) process results
are obtained.

It is unfortunate, but all too frequently true, that production
people are reluctant to receive unfavorable data from a new

process, or process change, and would like to shoot the messenger who brings it.

The laboratory, sad to relate, is the messenger.

Anything that can be done prior to start-up to allay this
suspicion is worthwhile doing.

VIII.   LABORATORY SAFETY

Many, if not most, chemical companies, today, have employee
safety programs. Safety in the laboratory is, or certainly should
be, generally included in such programs.

A chemical laboratory can be a safe place in which to work,
but, considering the ever present potential hazards of toxic gases,
poisons, open flames, acids, oxidizers, caustics and glassware,
among many others, it requires constant vigilance on the part of
the staff to keep it safe.

If your laboratory doesn't have a manual or book discussing
laboratory safety, the "Handbook of Laboratory Safety" 2nd Edition, published by CRC Press, Inc., 2000 N.W. 24 Street, Boca
Raton, FL 33431, is highly recommended.

I would suggest, as well, that you obtain a copy of OSHA
Safety and Health Standards (29CFR 1910), OSHA 2206, (Revised
November 7, 1978). This may be purchased from the Superindent
of Documents, U.S. Government Printing Office, Washington, D.C.
20402.

In the event your laboratory doesn't have an ongoing safety
program, let me suggest the following steps to develop one.

1.   A safety policy statement should be written, signed and
     issued by a top management official.

2.   A safety committee should be organized. In a small
     laboratory this can be one or two staff members, alter-
     nating on a predetermined basis.

3.   Rules and regulations applicable to laboratory safety
     should be written by the committee members.

4.   Safety inspections by the committee should be carried
     out on a regular basis to monitor conformance to the
     rules and regulations.

5.   Accidents, however small, should be investigated and
     recorded by the safety committee.

6. Records of occupational illnesses and injuries must be prepared and maintained. This is an OSHA requirement. Obtain a copy of "Recordkeeping Requirements Under the Occupational Safety and Health Act of 1970." This is available from the U.S. Department of Labor.

7. Conduct regularly scheduled safety meetings and training programs. These can be tailored to your specific needs.

In the event you may wish further information, the following references on laboratory safety may be helpful.

1. "Guide for Safety in the Chemical Laboratory," 2nd Edition, Chemical Manufacturers Association. Van Nostrand Reinhold Co. Inc., New York, NY 10010

2. "Handbook of Compressed Gases," Compressed Gas Association, 500 Fifth Avenue, New York, NY 10036

3. "Standard First Aid and Personal Safety," 2nd Edition, American National Red Cross, Doubleday and Co., Inc., Garden City, NY 11530

4. "Accident Prevention Manual for Industrial Operations," 7th Edition, National Safety Council, 444 N. Michigan Avenue, Chicago, IL 60611

SOURCES FOR ANALYTICAL METHODS

Welcher, P. J., "Standard Methods of Chemical Analysis", 6th ed., Van Nostrand, Princeton, New Jersey, 1962-1966.

Christian, G. D., "Analytical Chemistry", 2nd ed., John Wiley, New York, 1977.

Hillebrand, W. F. and Lundell, G. E., "Inorganic Analysis", 2nd ed., John Wiley, New York, 1953.

Siggia, S., "Quantitative Organic Analysis Via Functional Groups", 3rd ed., John Wiley, New York, 1972.

Ewing, G. W., "Instrumental Methods of Chemical Analysis", 4th ed., McGraw-Hill, New York, 1975.

Assn. of Official Analytical Chemists (formerly the Assn. of Official Agricultural Chemists), "Official Methods of Analysis of the AOAC", quinquennial, with annual supplements, AOAC, 1111 N. 19th St., Arlington, VA 22209.

U.S. Environmental Protection Agency, "Methods for Chemical Analysis of Water and Wastewater", U.S.E.P.A., Washington, D.C., 1979.

American Service for Testing and Materials, Annual Book of ASTM Standards, (revision issued annually), ASTM, 1916 Race St., Philadelphia, PA 19103.

National Institute of Occupational Safety and Health (NIOSH), "NIOSH Manual of Analytical Methods", Vols. 1-6, 2nd ed., U.S. Dept. of Health and Human Services, Washington, D.C.

Another source is the manuals of analytical procedure supplied by instrument manufacturers with purchase of their equipment.

3
SAMPLING

In my experience, probably the greatest hindrance to effective
quality control is inadequate sampling. This is especially true for
raw materials. If the sample is not representative of the raw material, it is virtually without value as an indicator of what is being
received.

It's true that there may be other causes for poor analytical
data having to do with laboratory techniques, reagents, instrumental problems, etc. These are all capable of being tracked
down, or compensated for, or corrected. Poor sampling procedures or techniques leading to invalid samples are something else
again.

In-process quality control is gaining favor in some quarters.
This is presumed to replace some part of the sampling done on the
finished product. In-process control utilizes on-line analytical
instrumentation, or taking samples from designated sampling points
in the process.

This does not, however, relieve you of the responsibility for
the quality of the finished product. There are thousands of small

or modest sized chemical plants which do not use, or cannot afford, on-line instrumentation, and must rely upon manual sampling of the end product.

First, let's look at some of the ways raw material may be sampled:

I.   RAW MATERIALS: LIQUIDS

In general, a plant will receive liquid raw materials in one of three ways: in discrete containers, such as drums; in tank cars or tank trucks; and via pipelines from a nearby plant.

Background knowledge of the liquid raw material can be important.

Does it have a tendency to layer out?

Is it liquid at lower than ambient temperatures?

Is there likely to be — or even, is there a possibility that there will be — crystal formation, sludge formation or precipitation?

Should the product be kept in a warming room prior to sampling or use to be certain it is a homogeneous liquid?

Should heating coils be used in tank cars or trucks to keep material in a liquid state?

Liquid raw materials can vary significantly in density, color and odor; they can vary from mild to corrosive; they can require the use of protective clothing and respirators while being handled.

Will a liquid have an effect on the vessel in which it is stored, or the process equipment through which it must pass? As basic as this knowledge appears to be, there are a surprising number of process units which have had to be rebuilt of different materials simply because there was either inaccurate knowledge, or no knowledge, or ignored knowledge of the effects of a liquid raw material on process equipment or storage facilities.

Has the liquid been transported from supplier to your plant in a lined or unlined vessel?

If shipped in an unlined vessel — tank car, tank truck, barge, whatever — the liquid could pick up iron or another contaminant. If these are undesirable in the finished product, it becomes obvious that not only must care be taken in the design or use of storage or process equipment but that specific shipping containers must be stipulated as well.

It can be seen that there is much necessary information to be obtained prior to sampling.

A. Discrete Containers

Assuming that this is the best of all possible worlds and that the questions posed above have been satisfactorily addressed, let us also assume that the liquid raw material has been delivered in a steel drum. What is a good way of sampling it?

Insert a wooden rod long enough to reach the bottom of the drum and feel for sludge, crystals or precipitate. Examine the bottom of the rod after it has been withdrawn and note whether the liquid is indeed homogeneous. Additionally, roll the drum on its side sufficiently to assure homogeniety of its contents.

A sample may then be withdrawn in any suitable fashion — i.e., by dipping, siphoning, pouring or insertion of a spigot — and then analyzed as appropriate.

B. Tank Cars or Trucks

Materials delivered in tank cars or tank trucks may or may not be homogeneous despite the shaking that occurs during transportation. The liquid may move as a solid mass in the tank and not really mix much.

Before sampling, if feasible, open the dome at the top of the tank, look in and note whether there may be flecks of tank lining floating on the liquid. This could be important if, for example, you were receiving a shipment of liquid caustic. If enough lining had been sloughed off, the iron content of the caustic could have been increased sufficiently to affect its use in your process.

After this, the tanks may be sampled from the bottom, top or side, depending on where the sample valve is located.

For a variety of reasons — including the volume involved, possible layering, or crystal formation — it may be desirable to use a multi-level sampling device. Multi-level samplers are not generally available as commercial items, and are usually fabricated in the plant shop, or otherwise custom made.

A multi-level sampler (Fig. 1) consists of a long support, carrying brackets to hold two to five sample bottles at different

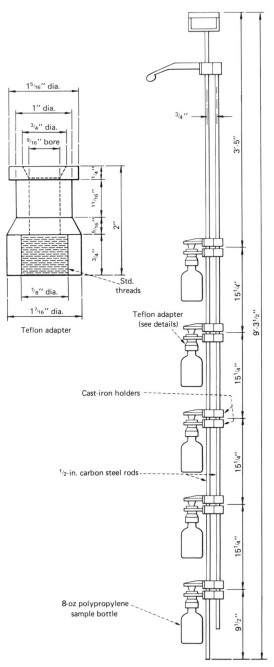

1⁵/₁₆″ dia.

1″ dia.

³/₄″ dia.

⁹/₁₆″ bore

¹/₄″

¹¹/₁₆″

⁵/₁₆″

³/₄″

2″

Std. threads

⁷/₈″ dia.

1⁷/₁₆″ dia.

Teflon adapter

³/₄″

3′-5″

Teflon adapter
(see details)

15¹/₄″

9′-3¹/₂″

Cast-iron holders

15¹/₄″

¹/₂-in. carbon steel rods

15¹/₄″

15¹/₄″

8-oz polypropylene
sample bottle

9¹/₂″

Figure 1. This five-level sampler can be made in the plant shop.
(Reprinted by special permission from CHEMICAL ENGINEERING,
April 7, 1980, by McGraw–Hill, Inc., New York, N.Y. 10020.)

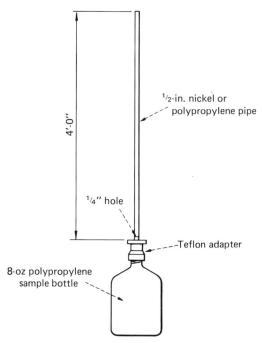

Figure 2. A one-level sampler for deep tanks. (Reprinted by special permission from CHEMICAL ENGINEERING, April 7, 1980, by McGraw-Hill, Inc., New York, N. Y. 10020.)

levels. Next to one support rod and extending its entire length is a rod having a cross-piece at the top to serve as an operating handle. Attached to this rod, and sliding on the other rod, are arms, each carrying a stopper for the corresponding bottle.

To secure samples, the handle is pressed down, closing the bottles. The sampler is then lowered to the bottom of the tank car, tank truck or storage tank, the handle raised, and the bottles allowed to fill. After bubbles stop rising, indicating that the bottles are full, the handle is depressed once again, closing the bottles.

The sampler can now be removed from the vessel and suitably cleaned before handling.

For single level sampling through the top, refer to the sampling devices shown in Fig. 2 and 3. Obviously, the choice of construction materials or bottles for these samples depends on the type of liquid being sampled.

Sampler, one-level
1-pint dipper
3, 6, 12-ft wood handles

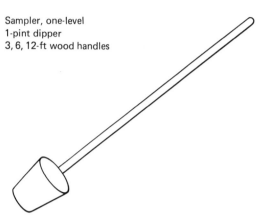

Figure 3. This one-level sampler is a simple dipper. (Reprinted by special permission from CHEMICAL ENGINEERING, April 7, 1980, by McGraw-Hill, Inc., New York, N.Y. 10020.)

It must be stressed, once more, that after sampling, the exterior of the bottles should be handled with care and washed or otherwise cleaned before removal of the sample. This will help prevent skin burns or irritations.

In the event that bulk receipts are transferred to storage or holding tanks prior to use, the material may be sampled through a sampling valve on the side of the storage tank. Liquid raw materials received via pipelines are usually pumped directly into storage tanks and may be sampled from those tanks.

In holding or storage tanks where layering or development of any sort of precipitate may be a problem, consider installation of an impeller, operating at low speed, or a gas or air bubbling system, to keep the liquid homogeneous.

II. RAW MATERIALS: SOLIDS

This type of raw material may be received in bulk, such as hopper cars, Tote Bins, or even large capacity fiber drums, or in kraft or poly bags or small drums of varying size and construction.

Sampling solids is not as clearcut an exercise as sampling liquids. For example, there may be problems with stratification due to difference in particle size or density; or an agglomeration of

particles; or dusting-breakdown of large particles. Obviously, if
any of these factors is present, getting a homogeneous sample is
a problem.

Thus, a grab sample from a hopper car or other bulk vessel
should not be considered representative of the container's contents and should not be used. It is best to use a "thief" to accom-
plish vertical sampling, top to bottom, and to repeat this procedure two to four times in different areas of the car.

A thief is a tubular device used to secure samples from different levels of containers such as cars, Tote Bins, and fiber
drums, as well as for sampling granular materials.

It consists of two polished brass telescoping sheeted tubes,
with slots of the inner and outer tubes registering so that they
may be opened and closed for collecting a sample by turning the
inner tube. The tubes, generally, are about 40 inches long with
a 1 3/8 inch diameter.

They are available from scientific apparatus supply houses
and are classified as sampling tubes.

III.   RAW MATERIALS: GASES

Whether a gas is delivered in a tank car or a cylinder, a grab
sample is acceptable. Basically, this is generally collected by
passing the gas through a cooling coil into a dewar flask (a dewar
flask is simply a temperature-maintaining container similar to a
thermos bottle. It has double walls, is internally silvered and
highly evacuated to prevent heat transfer between contents and
atmosphere.) A 4200 ml flask is desirable for this purpose.

IV.   FREQUENCY OF ANALYSIS

When raw materials are purchased for the first time from a new
vendor, every lot or batch should be completely analyzed on receipt for properties listed in the specification covering the product. Generally, this is done until five successive samples have
been shown to be satisfactory.

After that, the raw material may be sampled on a less frequent basis, with analysis of only the more important properties
necessary to describe the product and assure its quality. Don't
sample so infrequently that you lose touch with the quality of your
receipts.

Should it become known to you that there has been a process change or component substitution in the manufacture of the raw material, it will become necessary to revert to the original five sample plan noted above.

The same five sample plan would hold even if the supplier shifted the source of the raw material from one of his plants to another, certainly it would be time if the source of the raw material were shifted from one country to another.

This should be done whenever the quality of the raw material is found to be less than completely satisfactory for whatever reason.

Where practicable, samples of raw material should be retained, under suitable storage conditions, over the estimated useful life of the product in which they have been used.

V. PRODUCTION SAMPLING

The frequency of sampling for process or quality control is, by and large, a function of the process and its complexity.

For example, in a simple batch process, liquid or dry, sampling should be done once for each batch. Since each batch is subject to variation, however slight, every batch sample must be analyzed for conformance to specification.

A continuous process that shows only minor variability may be sampled once or twice per shift, at designated sample points, and analyzed for some particular property that will readily indicate whether or not the process is on target.

(Determination of process variability will be discussed in Chapter V under the heading of Control Charts.)

A complex continuous process should be sampled more often, perhaps every two hours. If the process demonstrates variability, in addition to its complexity, hourly samplings may be indicated.

On whatever basis the sampling frequency schedule is designed, it is most important that sufficient sampling, and analysis, be done to detect process upsets quickly and to be certain that manufactured product continues to meet specifications.

Since the essence of quality control is the validity of the sample, it would not be amiss here to discuss the subject of who actually does the sampling.

In most plants, production samples are taken by production personnel and shipment samples are taken by shipping personnel. But, where feasible — and I recognize it is not always practicable — laboratory personnel should participate in the sampling process.

In any event, whether feasible or not, quality control personnel should:

1. Train, or assist in training, plant personnel in the correct ways of sampling;

2. Prepare and issue sampling containers to them; and

3. Agree, mutually, upon sampling frequency and the delivery of samples to the laboratory on time.

An abrasive point of contention between production and quality control in many plants is the number of resamples to be taken when an initial analysis shows the product to be somewhat out of specification or even unacceptable.

It has never ceased to amaze (and amuse) me that production personnel will cheerfully accept analytical results which show their product to be satisfactory, but become enraged at the "incompetence" of the same laboratory analysts who have the temerity to tell them that their product is "out of spec" and demand resample after resample.

It is probably not stretching fact to say that if one were to sample enough, an acceptable sample could eventually be found. Unfortunately, such over-sampling is practiced more often than admitted. But to whatever extent it is done, it negates quality policy.

It should be a steadfast rule in your plant, that in the event of failure of the original production sample, for whatever reason, only one resampling will be permitted. Whenever possible, this one resampling should be performed by quality control personnel, or, at the very least, supervised by them.

A. Production Sampling: Liquids

Liquids must be thoroughly mixed, single-phase, with any solids completely dissolved. In general, liquids are the easiest to sample — either in the process stream, or as finished products.

Engineering or production personnel will normally have selected sampling points at which the process stream is homogeneous and will have installed sample take off valves at these points.

In a continuous liquid process, one need only determine the sampling frequency necessary for control, and choose a sample vessel that is adequate for the product and is non-contaminating in itself.

For liquid products manufactured by batch processing, one sample per batch is generally adequate. This will normally be taken from the sample tap on the batch processing tank.

In the case of liquids being loaded as finished product into tank cars or tank trucks, it is advisable to sample from the car or truck itself. This is what the customer will receive, and this is what should be tested. In the event of a contaminated tank car or truck (or barge), it will be this sample which can show it.

Similarly, liquid samples should be taken from drum shipments for the same reasons. In the event of a problem with a customer, these samples, kept as retains, will be more representative of what the customer received than samples taken from process lines or sample valves prior to filling or unloading.

B.   Production Sampling: Solids

Solids (crystalline, granular, or powdered products) are generally sampled in one of three ways:

1. A grab sample is taken several times a shift at the bagging machine (or drumming spout).

2. Grab samples are taken, on a random basis, from one or more containers. This is most easily accomplished with valve-type bags or with drums. With sewn bags, it is advisable to sample at the hopper. Such samples may be analyzed indirectly — or composited and run as one sample.

3. Manufactured material destined for bulk shipment may be sampled in the hopper car or truck, using a thief (this device has been described previously). A drawback to this mode of sampling is that the car or truck might be shipped to a customer before the laboratory is able to analyze the sample.

There are, of course, process samples which can be taken from evaporators, slurry tanks, driers, crystallizing beds, etc. The prime concern in any process sample is, of course, that it be representative of the stage of the process at which it is taken.

C. Production Sampling: Gases

Grab samples from gas production are satisfactory. These can be taken either by production personnel or by a laboratory technician directly from the tank car after loading. If, for any reason, this is impracticable, sampling may be done at the loading header as the tank car is being loaded.

As described before, the sample is collected by passing the gas through a cooling unit into a dewar flask (4200 ml). Any discussion of sampling would be incomplete without mention of U.S. Military Standard 105D "Sampling Procedures and Tables for Inspection by Attributes," dated 29 April 1963 (and not revised or reissued since).

The principal thrust of this standard is the development of Acceptance Sampling Programs through the use of sampling tables and Acceptable Quality Levels (AQL).

Table I. Sample-size code letters. (Reprinted by special permission from CHEMICAL ENGINEERING, June 16, 1980, by McGraw-Hill, Inc., New York, N.Y. 10020.)

Lot or batch size			Special inspection levels				General inspection levels		
			S-1	S-2	S-3	S-4	I	II	III
2	to	8	A	A	A	A	A	A	B
9	to	15	A	A	A	A	A	B	C
16	to	25	A	A	B	B	B	C	D
26	to	50	A	B	B	C	C	D	E
51	to	90	B	B	C	C	C	E	F
91	to	150	B	B	C	D	D	F	G
151	to	280	B	C	D	E	E	G	H
281	to	500	B	C	D	E	F	H	J
501	to	1,200	C	C	E	F	G	J	K
1,201	to	3,200	C	D	E	G	H	K	L
3,201	to	10,000	C	D	F	G	J	L	M
10,001	to	35,000	C	D	F	H	K	M	N
35,001	to	150,000	D	E	G	J	L	N	P
150,001	to	500,000	D	E	G	J	M	P	Q
500,001	and	over	D	E	H	K	N	Q	R

Table II. Single sampling plans for inspection (master table). (Reprinted by special permission from CHEMICAL ENGINEERING, June 16, 1980, by McGraw-Hill, Inc., New York, N.Y. 10020.)

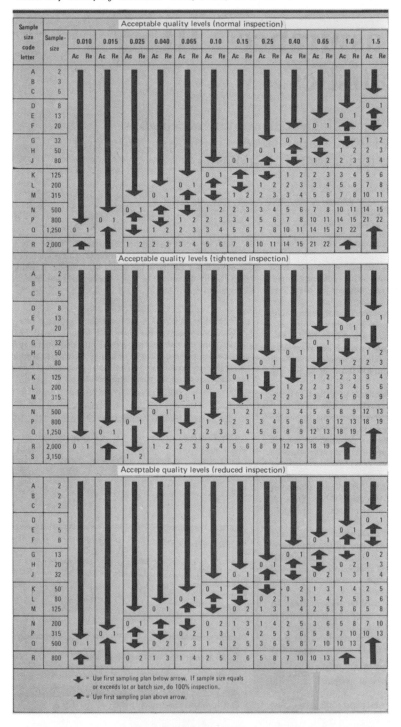

Table II. (continued)

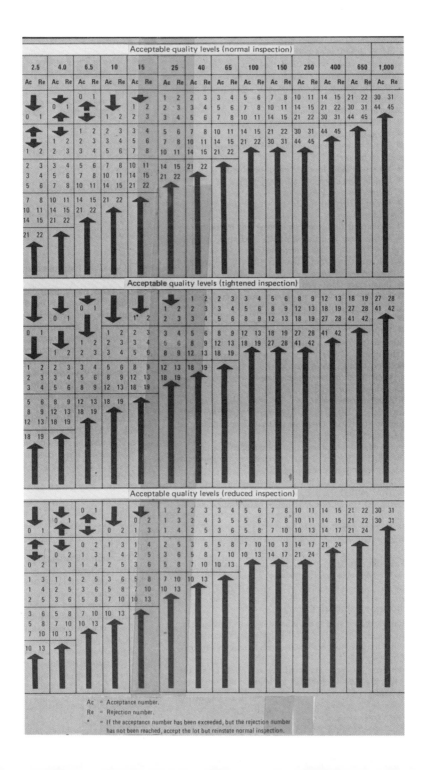

The AQL, which is explained in the standard, is the maximum percentage of defects that, for purposes of sampling inspection, can be considered satisfactory or acceptable. Defects are classified as critical, major or minor.

Originally designed for use with hardware sold to the military, it can be converted fairly easily to use in the chemical process industry.

Table I, taken from this standard, is used to determine inspection levels, which in turn determine the relationship between lot or batch size and the sample size. In general, Inspection Level II is the most commonly used. Inspection Level I may be used where less discrimination is needed, while Level III may be specified for greater discrimination.

The four special levels: S-1, S-2, S-3, S-4 noted in Table II are used where relatively small sample sizes are necessary and large sampling risks can or must be tolerated.

In the column - Lot or Batch Size - the number groupings may indicate empty containers, or filled containers, or pounds of production — or any discrete unit capable of being sampled.

Thus, if we were trying to sample a shipment of 5,000 empty containers, we would find the grouping 3,201 - 10,000. Proceeding along that line to Inspection Level II, we note the letter L.

In Table II, single sampling plans for inspection (master table) we find letter L under sample size code letter (first column) in the first table designated as normal inspection. Next to the letter L we find the sample size to be 200. This then, is the number of containers to be inspected for the particular attribute you need. If you have confidence in your supplier, and past shipments have been satisfactory, try letter L in the table for reduced inspection. Here you will find the sample size to be 80 containers.

When the normal inspection plan is being used, the switch to tightened inspection should be made when two out of five consecutive batches or lots have been rejected on original sampling.

Return to normal inspection is indicated when five consecutive lots or batches under tightened inspection are acceptable. Similarly, the switch from normal to reduced inspection may be made when ten consecutive lots or batches have been acceptable on original inspection.

After finding the sample size, read across the table to the appropriate AQL to find out how many defective samples (or defects) permit acceptance, or call for rejection of the lot or batch. Table III depicts in abstracted form, numbers applicable to an AQL plan, operating under normal inspection.

Table III. Sampling and inspection. (Reprinted by special permission from CHEMICAL ENGINEERING, June 16, 1980, by McGraw-Hill, Inc., New York, N.Y. 10020.)

MIL STD 105D , normal inspection, general level II

Lot size	Sample-size code	Sample size	Number of defects for acceptance (Ac) or rejection (Rj) at each AQL							
			0.065		0.65		1.5		4.0	
			Ac	Rj	Ac	Rj	Ac	Rj	Ac	Rj
501 – 1,200	J	80	0	1	1	2	3	4	7	8
1,201 – 3,200	K	125	0	1	2	3	5	6	10	11
3,201 – 10,000	L	200	0	1	3	4	7	8	14	15
10,001 – 35,000	M	315	0	1	5	6	10	11	21	22
35,001 – 150,000	N	500	1	2	7	8	14	15	21	22

To an extent, the selection of an AQL is subjective. Ask yourself, how many critical defects can we tolerate in a materials receipt (or produced lot) before it becomes essential to reject the entire shipment, or a specific lab or batch within a shipment?

Consider the same question for all major and minor defects. Select an AQL that most closely matches your answers to these questions. Examples of AQL's and critical, major and minor defects will be found in some detail in Chapter IV on Quality Control of Packaging Materials.

4
QUALITY CONTROL OF
PACKAGING MATERIALS

I. INTRODUCTION

Packaging materials represent a multi-billion dollar industry.
Many hundreds of millions of these dollars are spent by the chemical process industries to provide the packaging to move their products to their customers.

With the exception of the product moved in bulk or semi-bulk via pipelines, tank cars and tank trucks, tote bins and so called "Gaylords" virtually everything else is shipped in steel or fiber drums, multi-wall bags or plastic containers of one size or another.

To many, if not most customers, the image they have of your company is the condition and quality of the package in which your product arrives. If your container is poorly sized or designed; is rusted; dented; torn or bulged; or looks tired and shopworn; your company's image will suffer.

For your product to arrive at the customer in a condition to be utilized, a suitable container must be selected and its quality monitored.

41

The importance of selecting proper packaging must not be underestimated, and the assurance of the quality of these containers is vital to the safe and efficient transportation of your company's product.

Packaging is a technical function and should be performed by qualified personnel. In small companies this can usually be handled by laboratory personnel because of their knowledge of the product characteristics. Labelling and marking of all chemical packaging is a legal responsibility and should be delegated either to in-house counsel or an outside attorney familiar with pertinent regulations.

The transportation of hazardous materials and, indeed, most chemicals in the United States is governed by the Department of Transportation regulations under the authority assigned to it by the Transportation Safety Act of 1974. These regulations apply to air, water, rail, and highway transportation. Adherence to the U.S./DOT regulations is the first requirement of chemical packagers. The applicable DOT regulations are outlined in the Code of Federal Regulations, Title 49-Transportation, Parts 100 to 177 and 178 to 199, (1980) and copies of these documents are available from : Superintendent of Documents, U.S. Government Printing Office, Washington, D.C. 20042. These regulations take precedence when selecting the package.

II. CONTAINER FORMS

Chemical products fall into three categories: liquid, solid, or gaseous. The latter product is specialized and is transported in pressurized vessels. Most chemical process products are liquids or solids and it is to them we will focus our attention.

A. Liquids

1. Steel drum

The basic container of the industry is the fifty-five gallon tight head steel drum which is probably the most standardized industrial container in the world. Dimensions and locations of the openings are virtually the same from any supplier. You have a choice of

Figure 1.

2 1/4 inch or 3/4 inch threaded or plastic clench-on fittings. Any steel drum vendor can supply information and samples. Side bung openings and interior agitator types are also available for special emptying procedures or very viscous products.

The other common sizes in steel drums are fifteen and thirty gallon capacities. These can be ordered in a variety of diameters and heights to suit your needs. Standard diameters are 14, 18 1/4 and 22 1/2 inches. It is best however, to utilize standard sizes to maintain flexibility and availability of supply.

Essentially the containers are fabricated by rolling a sheet of steel into a cylinder and welding the edges. The heads or ends are then preformed with bung fittings already applied and then crimped to the body. A sealing compound is added to the area to be crimped prior to fabrication to insure a leak-proof container. Every drum is pressure tested with air before being certified for shipment. Unless otherwise specified by DOT requirements the gauge (thickness) of steel used is a function of the weight to be contained. For most liquids in a fifty-five gallon size the combination of twenty gauge steel bodies and eighteen gauge steel heads is adequate. If returnable service or multi-trip usage is desired, then sixteen gauge or heavier metal should be considered. Remember, the lower the gauge number, the heavier the steel. If a thirty gallon or fifteen gallon size is required, lighter gauges may be selected, because less product weight is to be transported.

For liquids incompatible with raw steel, for whatever reason, coatings can be applied to the inside of the drum after fabrication. Drum linings or coatings keep your product from reacting with the raw steel surface of the container and also prevent the formation of rust in the drum prior to filling. These are generally sprayed on and then cured by baking at elevated temperatures. When your supplier discusses linings with you, remember that a number of factors must be considered in the selection process: the product itself; the resistance of the lining to chemical reaction with the product; impact resistance of the lining (a cracked lining is no better than no lining at all), product shelf life; and, certainly, cost.

Test panels and test cups are generally available from your drum supplier to perform your own compatibility tests. Such tests are generally run at 125° to 140°F for about two weeks. This is roughly equivalent to about two months storage at ambient temperatures. While the drum vendors will test your product — no lining is ever guaranteed — it is up to you to be satisfied that the coating meets your requirements.

Figure 2.

2.   Composites

There are combinations utilizing inner high density polyethylene
drums with outside shells of steel. These are called composites
and a wide variety of open and tight head designs are available.
Most of these composites are appreciably more costly than lined
steel drums but can be considered "top of the line" for products
that are especially corrosive.

Composites are also available in fiber shells. The outer containers are fabricated similarly to the fiber drums described under
the solid product section following. This offers a less expensive
composite version than the steel drum.

3.   Plastic Drums and Carboys

An industrial container that is enjoying an increasingly larger
chemical market is the all plastic drum. The most popular sizes
are 15, 30 and 55 gallons. However, there are differences among
the configurations with respect to handling and design. The
drums are molded from high density polyethylene resin generally
pigmented black to reduce ultraviolet degradation, a phenomenon
that embrittles plastic after prolonged exposure to sunlight. Before selecting this type of package, make sure that the materials
handling equipment of your company, as well as that of your
customers, has the capability of moving these drums.

Plastic carboys are also available in a 15 gallon capacity and
are relatively high priced, but they generally use excellent closures.

Both heavy wall plastic drums and carboys are suitable for
returnable service. However, an in-house quality recheck program must be designed to cover cleanliness and physical integrity
of the drums and carboys prior to refilling and reshipping.

B.   Solid Materials

Depending on hazard and storage conditions, the best containers
are open head steel and fiber drums.

1.   Steel Drums

Diameters are limited, as in the tight head steel drums, to 14 inch,
18 1/4 inch, and 22 1/2 inches. Closures in general usage are
the external bolted ring type or external lever lock, and are

Figure 3.

fabricated from heavy gauge steel. In the smaller sizes and lighter gauges the lug type cover is available. An advantage to this type cover is that automated closing equipment is available for high production volumes.

2. Fiber Drums

These containers are fabricated from plies of Kraft paper that are glued together while being convolutely wound on a mandrel. For heavy duty service the bottom, which is formed from layers of paper incorporating a waterproof medium and suitable adhesive, is crimped to the body and reinforced by a metal chime. The top chime also incorporates a metal ring which accepts a metal or fiber cover. A lever lock external ring completes the closure. The various carrier regulations require a wall thickness of .090-.240 mils depending upon drum size. Side wall strength is measured

Figure 4.

in Mullen test units which approximate bursting strength in pounds per square inch.

An advantage to this type container lies in its versatility in accepting a fabricated internal protective lining as well as multiple moisture barrier plies in the sidewall. Also where moisture resistance is a vital criterion, the silicate adhesive commonly used to cement the plies could be replaced with a hot melt resinous type which is water resistant. Fiber drums are available with such linings as:

Polyethylene

Aluminum/Polyethylene

Wax Coatings (a variety of formulations)

Polyester

Polyvinyldine Chloride

Kraft Paper/Aluminum Kraft

Any of the above can be incorporated in the sidewall when laminated to a Kraft paper ply.

Aluminum foil is probably the best barrier lining with respect to economy. It is effective against moisture and greases. If needed, such lining constructions can also be inserted into the drum bottom. If the interior lining route is selected, then, obviously, the bottom would require a lining, as well. However, the bottom lining need not necessarily be the same as the body lining.

Another version is an all fiber unit comprising an inner and outer body tube with the outer tube cut shorter than the inner to accept a fiber telescopic type cover. With the cover in place, adhesive sealing tape can be applied to the smooth surface to seal the container.

All of these fiber containers are available in a wide range of diameters and heights with 3 1/2 to 75 gallon capacity. This differs from steel open head containers which are to a large extent, dimensionally limited.

3. Paper Shipping Sacks

For solid materials, a multi-wall sack or bag is the most efficient and least costly package per pound of chemical product. There

are non-rigid or flexible packages made from 3 to 6 plies of paper
of various weights including specialized vapor barrier or strength
contributing plies. A major advantage is that a bag can be designed exactly to meet your compatibility, dimensional, and handling requirements.

When referring to paper the term basis weight is used and a
paper ply is usually between 40-70 lbs. This is the weight of a
ream of 500 sheets of paper of standard dimensions (3,000 sq. ft.).
Paper used in bags must meet certain tear and tensile strengths.
The most widely referred to documents for paper tests are Uniform
Freight Classification and Federal Specifications UUS48. Commonly used paper plies are designated as follows:

Natural Kraft - Untreated paper

Extensible Kraft - Same as natural but paper has more
strength

Natural Wet Strength Kraft - Paper is treated with melamine
resin to provide moisture resistance.

Extensible Wet Strength Kraft - Same as above but utilizing
extensible Kraft

Bleached Kraft - Paper which is white or colored. Most often
used as outermost ply of a bag for aesthetic and printability
reasons

Vapor barrier and protective inner plies are available. Glassine
papers and certain plastic laminations provide protection against
grease.

Moisture protection can be provided by polyethylene as a
coating or as a free film. High density polyethylene, polyvinyli-
dine chloride, polypropylene, etc. can be used for chemical,
moisture and gas permeability protection.

There are two general filling categories in bags; the first is
the "Open Mouth", where filling is done either by hand, gravity,
or compression. Closure by sewing, adhesion or heat sealing is
done at the plant site. The second is the "Valve." These are
filled through a tube, generally fed by augers or vibrators. *Internal*, self-sealing bag valves are available which close when bag
is inverted. Here the seal is induced by the weight of product in
the bag. *External*, tuck-in Kraft sleeves are folded and tucked
in manually after filling and when lined with polyethylene can be
heat sealed if so desired.

The above basic designs are manufactured utilizing a variety of sewing and adhesion techniques. Each has its advantages and disadvantages and an excellent primer on this package category is the PSSMA General Manual No. 6 which is available from the Paper Shipping Sack Manufacturers Association, Inc., 2 Overhill Road, Scarsdale, New York, 10583.

4. Bottles

Except for pharmaceuticals and certain laboratory chemicals which are packaged in glass, most smaller quantities of industrial chemicals are shipped in plastic bottles generally made of polyethylene. The advantages over glass are obvious. They are unbreakable, have less weight and require less of an overpack to ship. Standard liquid sizes are available with closure diameter sizes up to 38mm. Wide mouth bottles for solid materials have openings up to 120mm, and capacities of about one gallon.

The above are called "stock" bottles which means they are available from distributors in low volume quantities. If reasonably large volumes (100,000 and up) are contemplated, the design and fabrication of a private mold (your property) should be considered. This offers the advantage of designing the container to fit your filling and marketing requirements exactly. Initial mold costs can be appreciable ($5 - $50,000 depending on quantity and complexity). The economics should certainly be considered before committing to this type of package.

The closure is a vital part of this package and here again there is a wide variety available ranging from metal to plastic. Generally, the plastic bottle suppliers can and will make recommendations. However, the decision for final fit and degree of hermeticity required is your responsibility and should be thoroughly researched and tested in your plant. Remember that any package is only as efficient as its closure.

5. Corrugated Boxes

These containers are somewhat similar to bags in the sense that they offer flexibility of design. Boxes are generally fabricated by passing a 26 pound basis weight Kraft paper over steaming metal rolls which are grooved. The sheet assumes the configuration of the roll. This corrugated (fluted) Kraft media is glued between two sheets of 42 pound (outer) and 33 pound (inner) basis weight Kraft in large rolls and then converted to boxes by dye cutting and glueing. In the trade, terms like corrugated, cardboard,

fiber, etc. are used synonymously. There is some fabrication of
"solid" fiber boxes which is made by compressing wood fibers into
a solid sheet. These units are generally most costly. The "bag
in box" concept is widely used for a variety of non-hazardous
chemicals. This incorporates a polyethylene bag filled with product; sealed or tied and overpacked in a corrugated box. The
overpack can be designed to have its own telescopic cover similar
to one of the fiber drum designs but it is much more economical.
The sizes and weights should be limited to safe handling practices.
Corrugated boxes are also widely used for the shipment of plastic
and glass bottles. The separators or corrugated partitions used
in packaging glass bottles are for cushioning purposes to prevent
breakage. The glass bottles themselves help contribute to the
stacking strength. However, with plastic units stronger separators are required because plastic bottles do not contribute to
stacking strength because of their inherent flexibility. With plastic bottles the height of the separator should be slightly greater
than the height of the bottles to prevent any weight placed on top
of the box from crushing the bottles.

A handy method of receiving bottles and case has the corrugated box already assembled and the bottles placed in the case up-
side down to maintain cleanliness. This provides assembled packaging for the small plant. These units are called "reshippers".
This technique eliminates assembly on the part of plant personnel
and, of course, commands a slightly higher price from the supplier. The economics of assembling cartons in your plant should
be evaluated before purchasing this concept. The advantage of
assembly in your own plant is that less warehouse space is required for the unassembled units. Plastic bottles are then generally shipped in bulk with up to 200 bottles in one very large carton.

Just as it is sound practice to develop specifications for your
manufactured chemical product, it is equally important to develop
specifications for all of your packaging materials.

Several typical specification sheets for a variety of packaging
materials are shown in Figures 5-11. These will show you the sort
of information described by the property headings.

Much of this information can be developed by you with the
help of your supplier, or even supplied to you completely by your
vendor, in the event you are unable to determine some parameters
on your own.

In any case, it is mandatory to have the concurrence of your
supplier in any of the packaging component properties.

PACKAGE COMPONENT SPECIFICATIONS

Date Issued	Product	Code Number

Description

Steel Pail - 5 Gallons

PLANT:

DIMENSIONS: 11 7/16" dia. x 13 9/16" height overall

TARE WEIGHT: 5.0 lbs. approx.

SHIPPING SPACE: 0.980 cubic foot

VOLUME: 5.20 gallon minimum

CONSTRUCTION: Fabricated from 24 gauge throughout.
                     One 2" steel bung with polyethylene gas-
                     kets. Carrying handle in center of top.
                     Painted black. Interior to be clean and
                     free of dirt and foreign material. Heads
                     double seamed to body. Welded side
                     seam.

COMPLIANCE: To comply in all respects with DOT-17E.

PRODUCT LIST:

5 Gal. D.O.T. 17E Pail	Product Weight	Label
Product #1	50#	371 D.O.T. Flammable Liq.
Product #2	35#	270 D.O.T. Flammable Liq.
Product #3	50#	971
Product #4	45#	1272 D.O.T. Flammable Liq.
Product #5	40#	870 D.O.T. Flammable Liq.
Product #6	45#	1171 D.O.T. Flammable Liq.
Product #7	45#	1171 D.O.T. Flammable Liq.
Product #8	50#	370 D.O.T. Flammable Liq.
Product #9	40#	270 D.O.T. Flammable Liq.
Product #10	40#	370 D.O.T. Flammable Liq.
Product #11	50#	373 D.O.T. Flammable Liq.
Product #12	40#	870 D.O.T. Flammable Liq.
Product #13	50#	375 D.O.T. Flammable Liq.
Product #14	40#	769 D.O.T. Flammable Liq.

Figure 5.

PACKAGE COMPONENT SPECIFICATIONS

Date Issued	Product	Code Number
Description		

600 Lb. Plastic Drums

TYPE:	Tight head. Oval, rectangular
DIMENSIONS:	23" x 23" x 33"
TARE WEIGHT:	30 lbs.
CUBE:	10.10 cu. ft.
CAPACITY:	55 gallons
CONSTRUCTION:	Molded from high density, high molecular weight polyethylene with added block pigment. Drums to have 3/4" and 2" bungs with buttress threads.
COMPLIANCE:	To comply in all respects with DOT-34 specification.

Figure 6.

PACKAGE COMPONENT SPECIFICATIONS

Date Issued	Product	Code Number

Description	
	100 Lb. Fiber Drum

DIMENSIONS:	Outside - 14 7/8" Diam. x 20½" Height Inside - 14" Diam. x 19½" Height
NET WEIGHT:	100 lbs.
TARE WEIGHT:	6.8 lbs.
SHIPPING SPACE:	2.73 ft.^{3}
COMPLIANCE:	To comply in all respects with UFC Rule #51 and Section 173.245 b (a) (5) of DOT Regulations.
CONSTRUCTION:	7 plies of .012" Kraft Convolutely wound with inside ply of .002" PE coated on . .00035" of aluminum. Test wall strength 700 lbs. Mullen. Bottom consists of inner ply same as above laminated to Kraft. Total thickness of bottom .120" and test strength of 1000 lbs. Mullen.
COVER:	26 gauge lacquered steel w/rubber gasket and electrogalvanized locking ring. Cover to be raised for stacking.
PRINTING:	Plain
LABELING:	DOT Corrosive Material label to be applied. Add precautionary labels as required.

Figure 7.

PACKAGE COMPONENT SPECIFICATIONS

Date Issued	Product	Code Number

Description

55 Gallon Nominal (57 Gallon Actual) Removable Head Steel Drum

DIMENSIONS: Outside: 23 27/32" dia. x 34 13/16" high
                      Inside: 22½" dia. x 34 3/4" high

WEIGHT: Approximately 54 lbs.

CAPACITY: 7.63 cu. ft.

SHIPPING SPACE: 11.667 cu. ft.

MATERIAL: Low carbon steel - 18 ga. body and heads

COMPLIANCE: To comply in all respects with UFC Rule
                      40.

FINISH: Exterior: Body - #57 Black
                                 Top Head - #41 White
                                 Bottom Head - #26 Brown

                      Interior: Unlined - Phosphatized

CLOSURES: Size and No.: One each 2" and 3/4"
                                     diametrically opposed
                                     opening in top head
                      Flanges: Unplated steel, phospha-
                                     tized or equivalent
                      Plugs: same as flanges
                      Gaskets: Polyethylene
                      Cover: Union Cord. Code #5.5 -
                                     UC or equivalent
                      Protection: Cap seals
                      Locking Ring: Steel, 16 ga.

CONSTRUCTION: Drums shall be constructed in full compli-
                      ance with Steel Shipping Container Insti-
                      tute and Consolidated Freight Classification
                      Specifications.

Figure 8.

PACKAGE COMPONENT SPECIFICATIONS

Date Issued	Product	Code Number

Description
100 Lb. Multi-Wall Bag

TYPE:	Sewn Open Mouth
NUMBER OF PLIES:	4
TUBE SIZE:	18" x 3" x 36½"
CONSTRUCTION:	2/50 Natural Kraft, 1/.0005" HDPE, 2/50 Natural Kraft, Cotton filler cord, self-forming gussets. Spot paste all plies. Outer ply to have non-skid coating applied.
CLOSURE (MGG. END):	Machine-sewn through 70 lb. 2 1/8" natural tape with 5/12 colored looper thread, regular reinforcing cord. 2 mil. Thermogrip.
PRINTING:	Plain - Show code number, construction and date code under certification of bag maker.
COMPLIANCE:	To comply in all respects with UFC Rule #40.
TARE WEIGHT:	1.0 lbs.
CLOSURE (AT PLANT):	machine-sewn through 2½", 70 lb. natural creped tape with 6/12 white cotton sewing thread.
PACKED:	Per manufacturer's standard on 4-way entry pallet. All pallets to be stenciled on four sides with purchase order and code numbers and bag quantity per pallet.

Figure 9.

PACKAGE COMPONENT SPECIFICATIONS

Date Issued	Product	Code Number

Description

Plastic Bottle

STYLE:	Tapered wall cylinder
DIMENSIONS:	Height overall 14 9/16" ± 3/64 Diameter 5 11/16" (max.)
CAPACITY:	4,258 c.c. ± 60 c.c. to overflow (144 fl. oz. nom.) 24 hours after surface treatment and one pass through oven.
WEIGHT:	150 grams ± 5 grams
MATERIAL:	Bleach grade high density polyethylene (blue-white).
FINISH:	72 mm. finish to accept closure manufactured by XYZ Plastics Corp. in accordance with Print #C-2211-4 dated 1/31/78.
SPECIAL REQUIREMENTS:	Bottles to be clean and free from oil, grease, organic material, solvent and dirt and subject to AQL standards. Minimum wall thickness .035". Bottles to be packed in cartons neck down. Carton bottom flaps to be glued.
LABELING:	Labels to pass gummed tape adherance test.

Figure 10.

PACKAGE COMPONENT SPECIFICATIONS

Date Issued	Product	Code Number
Description		

Corrugated Carton - 12 Bottles

STYLE: RSC

DIMENSIONS: Interior: 13 5/16" x 9 15/16" x 9 1/4"

CONSTRUCTION: Natural outer kraft. Body formed from
                    200# C-flute corrugated board. Glued
                    manufacturer's joint.

                    Partitions: Partitions from 200# C-flute
                                 corrugated board.

                                 (2) 9 3/16" x 13 1/4"
                                 (3) 9 3/16" x 9 7/8"

                    Partitions slotted and assembled to form
                    12 equal cells.

COMPLIANCE: To comply in all respects with UFC Rule
                    41.

PRINTING: Format to be submitted as required.
                    Colors to be in accordance with GCMI
                    color guide.

PACKING: Bundled in accordance with normal com-
                    mercial practice.

Figure 11.

III.   ACCEPTABLE QUALITY LEVELS (AQL's)

In this less than perfect world, it may be expected that from time to time packaging materials, as delivered to you, may not meet your specification.   To determine this, and, to determine how many off-specification units you may reasonably (and safely) accept, it is desirable to formulate an Acceptance Sampling Plan.

Such a plan is generally described in terms of Acceptable Quality Levels (AQL's), the basis for which is found in Military Standard 105D "Sampling Procedures and Tables for Inspection by Attributes," dated April 29, 1963.   There has been no more recent revision.

Generally, the Single Sampling Plan is used for Normal Inspection at General Inspection Level II.

For inspection purposes, a lot may be defined as one truckload or one carload of containers - or any other packaging component.   The unit of inspection should be one container (or one carton, one bottle cap, etc.).

The number of units inspected should correspond to the sample size indicated in Table II (from Chapter 3) "Sample Size Code Letters" of Military Standard 105D.   Sample units must be randomly selected.

If the number of defectives found in the sample is equal to or greater than the rejection number for the class of defects being considered, the entire lot is to be rejected.   (In Table I from Chapter 3).

The choice of Acceptable Quality Levels, to an extent, is subjective.   How many defective units of a particular classification can you tolerate before shipping the entire lot back to your supplier?   Bear in mind that some of these defective units, if accepted, will find their way to your customers.   How will they react?

Acceptance Sampling Programs can be developed in considerable detail, which obviously, necessitates a considerable amount of inspection; or they can be developed in abbreviated form, touching only the high spots, as it were, and thus, necessitating less inspection time.   Both programs obviously have their virtues and faults.

There are examples of both full and abbreviated AQL plans for several types of packaging materials shown in the following pages.

It may be appropriate to remind you that all testing, wherever possible, should be of a non-destructive nature.   Fifty-five gallon steel drums are too expensive to cut the heads off even a

few so that drum linings may be inspected. Use a light source
through the two inch bung opening, instead.

IV.   SUPPLIERS QUALITY CONTROL

Your suppliers are, without question, best qualified to control the quality of the product they deliver to you to meet your
specifications. They are, in a sense, an extension of your production operations.

Your supplier's quality control program for the packaging
components he ships to you should be discussed with him and,
where feasible, monitored from time to time. Such auditing may
enable you to decrease the volume of inspection you do on his product in your plant.

We recognize there are limits to this. You may be located
many hundreds of miles from your suppliers; or in the case of
plants outside the United States, using U.S. suppliers for materials not available locally. The only contact with the supplier may
be via mail or the occasional visit by the supplier's local representative.

V.   ACCEPTABLE QUALITY LEVELS (AQL) AND DEFINITION OF
CONTAINERS

A.   AQL Plan—Steel Drums

1.   Critical Defects—0.065 AQL

Any defect which will prevent the use of the steel drum in its intended manner or which will cause the steel drum to react adversely with the product or which will cause the steel drum to malfunction for consumer use with regard to dispensing and closure.

   a.   Holes or openings of any size in the sides or bottom of
        the steel drum or in the covers which will prevent the
        container from holding the product.

   b.   Internal contamination - foreign material, dirt, grease,
        etc.

   c.   Locking rings which do not fit.

2. *Major Defects—0.65 AQL*

Any defect other than critical which may seriously affect the filling line operation or end use of the steel drum container.

    a. Bent or dented covers sufficient to cause problems in
        applying the covers.

    b. Drums incorrectly labelled.

    c. No gasket in covers.

    d. Edges of any metal jagged and likely to cause harm to
        users.

    e. Unreadable labels in cautionary copy area.

    f. Bottom crimped in such a manner that drums do not stand
        straight.

3. *Major Defects—1.5 AQL*

    a. Incomplete unit - cover or locking ring missing.

    b. Major dents in steel drum sidewalls seriously detracting
        from container appearance.

    c. Locking rings difficult to remove from steel drum - or
        difficult to reapply after drum filling.

    d. Jagged welds, dirty welds.

    e. Improperly gasketed covers, i.e. gasketing material
        improperly positioned in cover or gasketing material
        off-specification.

    f. Rolling hoops are uneven, not parallel to bottom of steel
        drum.

    g. Steel drum painting not per specification.

    h. Inside coating marred or scratched.

    i. Steel drums rusty.

    j. Excessive scratching - outside or inside.

    k. Incorrect ink color.

4.   Minor Defects—4. 0 AQL

Any defect which is a departure from established standards, having no significant affect on the container function, or which aesthetically is undesirable.

  a.   Minor dents in steel drum sidewalls not detracting from steel drum appearance.

  b.   Locking ring locks which are difficult to open or to close.

  c.   Any non-readable print in other than cautionary copy. No smears or mars.

B.   AQL Plan—Fiber Drums

1.   Critical Defects—0. 065 AQL

Any defect which will prevent the use of the fiber drum in its intended manner or which will cause the fiber drum to react adversely with the product or which will cause the fiber drum to malfunction for consumer use with regard to dispensing and closure.

  a.   Holes or openings of any size in the sides or bottoms of the fiber containers or in the covers which will prevent the containers from holding the product.

  b.   Internal contamination - foreign material, grease, dirt, etc.

  c.   Improperly heat-sealed seams such that seams are open.

  d.   Locking rings which do not fit.

  e.   Top or bottom chimes or cover not hot-dipped galvanized.

  f.   Tears or cuts in inner liner of container.

  g.   Grease or oil on top chime or cover. Chimes must be wiped clean after application.

2.   Major Defects—0. 65 AQL

Any defect other than critical which may seriously affect the filling line operation or end use of the fiber drum container.

    a. Covers bent or dented sufficiently to cause problems in applying the covers.

    b. Incorrect label on fiber drum container.

    c. Chimes applied too loosely or too tightly.

    d. No gaskets in covers.

    e. Edges of any metal components jagged and likely to cause harm to users.

    f. Locking ring locks which cannot be closed.

    g. Unreadable labels in cautionary copy area.

3. *Major Defects—1.5 AQL*

    a. Incomplete unit - cover or locking ring missing.

    b. Major dents in fiber drum sidewalls or chimes seriously detracting from container appearance.

    c. Soft spots in fiber drum sidewalls - due to delaminated plies or improper drying.

    d. Inner liner seam opening greater than 1/8-inch.

    e. Locking rings difficult to remove from fiber drum container - or difficult to reapply after drum filling.

    f. Jagged welds, dirty welds.

    g. Improperly gasketed covers, i.e. gasketing material improperly positioned in cover or gasketing material off-specification.

    h. Torn labels, labels which do not overlap properly, i.e., open label seams, labels improperly positioned, labels seriously abraded.

4. *Minor Defects—4.0 AQL*

Any defect which is a departure from established standards, having no significant affect on the container function, or which is aesthetically undesirable.

    a. No date of manufacture on bottom of fiber drum container.

    b. Minor dents in fiber drum sidewalls or chimes not detracting from container appearance.

    c.   Locking ring locks which are difficult to open or to close.

    d.   Dirt on labels not easily removed by wiping, waterstained labels, blistered labels, minor abrasions on labels.

    e.   Any non-readable print in other than cautionary copy.

    f.   No DOT stencil on container.

C.   AQL Plan—Plastic Pails

1.   Critical Defects—0.065 AQL

Any defect which will prevent the use of the container in its intended manner or which will cause the container to react adversely with the product.

    a.   Holes or openings of any size which will prevent the container from holding the product.

    b.   Internal contamination - foreign material, grease, dirt, etc.

2.   Major Defects—0.65 AQL

Any defect other than critical which may seriously affect the filling line operation or end use.

    a.   Finish damaged or molded improperly to the extent it will not permit cover application.

    b.   Finish dimensions sufficiently out of specification so as not to permit cover application.

    c.   Wall thickness below specification.

    d.   Any non-readable print in cautionary copy.

    e.   Covers do not have a gasket.

3.   Major Defects—1.5 AQL

    a.   Covers can be put on plastic pail but with difficulty, i.e., they drag or bind.

    b.   Covers fit on plastic pail too loosely.

    c.   Labels are incorrect.

    d. Labels are badly skewed.

    e. Labels have excessive bubbles, are not firmly adhered, are improperly positioned, are smeared with dirt, oil, grease, etc.

    f. Pails are out of round.

    g. Handles are missing from pails.

    h. Covers are badly dented sufficiently to cause problems in application. Fins on covers are bent.

    i. Plastic pails have dents.

    j. Plastic has insufficient pigment in it - pails are translucent.

    k. Holes for wire tag do not line up properly.

    l. Poor label adhesion - more than 10% of Dyna-Cal label is removed in Scotch Tape Test.

4. Minor Defects—4.0 AQL

Any defect which is a departure from established standards, having no significant affect on the container function or end product useability, or which is aesthetically undesirable.

    a. Color drift outside of appropriate color range - per individual plastic pail specification.

    b. Exterior dirt, grease, or dust not easily removed by wiping.

    c. Excessive scratching or scuffing.

    d. Labels applied skewed sufficiently to detract from appearance of plastic pail.

D. AQL Plan—Plastic Containers

1. Critical Defects—0.065 AQL

Any defect which will prevent the use of the container in its intended manner or which will cause the container to react adversely with the product.

    a.   Holes or openings of any size which will prevent the container from holding the product.

    b.   Internal contamination - foreign material, grease, dirt, etc.

2.   *Major Defects—0.65 AQL*

Any defect other than critical which may seriously affect the filling line operation or end use.

    a.   Finish damaged or molded improperly to the extent which will not permit closure application. Where applicable, child resistant closures must function properly, i.e., line up with lugs and snap in place.

    b.   Finish dimensions out-of-specification sufficiently to not permit closure application.

    c.   Wall thickness below that specified on individual bottle specifications. Improper wall thickness distribution sufficient to cause filling line problems.

    d.   Any non-readable print in cautionary copy.

    e.   Bottles in incorrect cartons.

3.   *Major Defects—1.5 AQL*

    a.   Primary dimensions out of specified tolerance range as indicated on appropriate article and finish drawings.

        1) Overall height.

        2) Major diameter.

        3) Finish dimensions sufficient to cause capping problems. Pinholes in locking lugs on child-resistant closures (where applicable).

        4) Capacity below that specified on individual bottle specification.

    b.   Tails left on containers.

    c.   Bottle not adequately flame treated.

    d.   Rocker bottoms which cannot be corrected with a 0.025 inch shim.

e. Vertical load resistance below that specified on individual bottle specification.

f. Gram weight below that specified on individual bottle specification.

g. Improperly assembled or improperly glued cartons.

h. Poor label adhesion - more than 10% of decorated area is removed in Scotch Tape Test.

4. Minor Defects—4.0 AQL

Any defect which is departure from established standards, having no significant affect on the container function or end product useability, or which is aesthetically undesirable.

a. Color drift outside of appropriate color range - per individual bottle specifications.

b. Exterior dirt, grease, or dust not easily removed by wiping.

c. Body flash exceeding 0.016 inch or tail flash exceed 0.047 inch.

d. Excessive scuffing or scratching.

e. Specifications and secondary dimensions as indicated on appropriate article drawing.

f. Dirt or grease on cartons.

g. Any non-readable print in other than cautionary copy.

h. Labels applied skewed sufficiently to detract from appearance of container.

E. AQL Plan—Reshipper Cartons

1. Functional A Defects—6.5 AQL

A functional "A" carton defect is one which causes the reshipper carton to be unable to contain the item for which it was intended, react adversely with the product, malfunction in shipping with regard to handling stability and closure, or be non-machinable on a packaging line because of dimensional deviations.

a. Loose liner facings - blisters, wrinkles, or delamination - on any part of a vertical wall involving three (3) or more flutes.

b. Delamination of liner facings on flap extending in more than 1/2-inch from flap end and running across ten (10) or more flutes.

c. Board measures under specified caliper.

d. Spliced facings or medium in vertical walls.

e. Facings and medium are mis-aligned by 1/16-inch or more on vertical walls.

f. Manufacturer's joint completely unglued for one (1) inch or more along length of joint, or extending more than 1/8-inch beyond the center of top or bottom score line, or adjacent panel extending beyond score of manufacturer's joint is glued, which may vary 1/8-inch.

g. Score, cut, or other dimensional measurement, more than 1/16-inch from specified dimensions - except box panel to which manufacturer's joint is glued, which may vary 1/8-inch.

h. Reshipper cartons are wet.

i. Warp of 1/4-inch in any twelve (12) inch or less sections on single wall.

j. Ragged edge cutting, extending in 1/8-inch and covering three or more flutes, or present so one cut extends over another and both involving three (3) or more flutes.

k. Partitions improperly folded or assembled.

l. Bottom flaps not sealed at one or more normally glued areas.

m. Containers do not fall easily out of carton, i.e., fit too tightly for easy removal of containers.

n. Incorrect partition height.

o. Partitions fall out of cartons when containers dumped.

2. *Functional B Defects—10.0 AQL*

a. Print crushed more than 0.015 inch.

b. Flaps overlap *or* gap more than 1/8-inch at open portion of box.

c. Lack of proper scoring.

d. Dirty or smudged boxes on a normally visible surface.

e. Extraneous scores along flute direction or pull roll marks along flutes when their presence interferes with fit or function on the packaging line.

f. Loose liner at cut flap - delamination not more than 1/2-inch extending along ten or more flutes *or* blisters in flap area.

g. Incomplete coverage of printing, incomplete copy, undecipherable copy, incorrect copy.

3. *Major Defects—15.0 AQL*

a. Print poorly registered on panels of boxes so that print extends into a score or cut, or is out of symmetry by 3/4-inch or more.

b. Flaps sealed so that they are askew by 3/8-inch or more.

c. Interior of reshipper carton contains dirt resulting in visibly dirty containers inside or outside.

d. Small gaps in printing, color shade variation.

VI. ABBREVIATED INSPECTION PROGRAM FOR IN-COMING INSPECTION

A. AQL and Definition of Defects of Fiber Drums

1. *Critical Defects—0.065 AQL*

Any defect which will prevent the use of the fiber drum in its intended manner or which will cause the fiber drum to react adversely with the product or which will cause the fiber drum to malfunction for consumer use with regard to dispensing and closure.

a. Holes or openings of any size in the sides or bottoms of the fiber containers or in the covers which will prevent the container from holding the product.

   b.   Internal contamination - foreign material, grease, dirt,
        etc.

   c.   Improperly heat-sealed seams such that seams are open.

   d.   Top or bottom chimes or covers not hot-dipped galva-
        nized.

   e.   Tears or cuts in inner liner of container.

   f.   Grease or oil on top chime or cover. Chimes must be
        wiped clean after application.

2.   *Major Defects—0.65 AQL*

Any defect other than critical which may seriously affect the
filling line operation or end-use of the fiber drum container.

   a.   Bent or dented covers sufficient to cause problems in
        applying the covers.

   b.   Incorrect label on fiber drum container.

   c.   Chimes applied to loosely or too tightly.

   d.   No gaskets in covers.

   e.   Edges of any metal components jagged and likely to cause
        harm to users.

   f.   Unreadable labels in cautionary copy area.

3.   *Major Defects—1.5 AQL*

   a.   Major dents in fiber drum sidewalls or chimes seriously
        detracting from container appearance.

   b.   Torn labels, labels which do not overlap properly, i.e.,
        open label seams, labels improperly positioned, labels
        seriously abraded.

B.   AQL and Definition of Defects of Steel Drums

1.   *Critical Defects—0.065 AQL*

Any defect which will prevent the use of the steel drum in its
intended manner or which will cause the steel drum to react

adversely with the product or which will cause the steel drum to malfunction for consumer use with regard to dispensing and closure.

    a.   Holes or openings of any size in the sides or bottom of the steel drum or in the cover which will prevent the container from holding the product.

    b.   Internal contamination - foreign material, dirt, grease, etc.

    c.   Locking rings which do not fit.

2.   *Major Defects—0.65 AQL*

Any defect other than critical which may seriously affect the filling line operation or end use of the steel drum container.

    a.   Bent or dented covers sufficient to cause problems in applying the covers.

    b.   Drums incorrectly labeled.

    c.   No gasket in covers.

    d.   Edges of any metal jagged and likely to cause harm to users.

    e.   Unreadable labels in cautionary copy area.

    f.   Bottom crimped in such a manner that drums do not stand upright.

3.   *Major Defects—1.5 AQL*

    a.   Incomplete unit - cover or locking ring missing.

    b.   Major dents in steel drum sidewalls seriously detracting from container appearance.

    c.   Steel drum painting not per specification.

    d.   Inside coating marred or scratched.

    e.   Excessive scratching - outside or inside.

    f.   Incorrect ink color.

VII.   AQL PLAN–PAPER LABELS

A.   Acceptable Quality Levels and Definition of Defects

1.   *Major Functional Defects–0. 65 AQL*

A major functional defect is one which causes the label (1) not to fit the container for which it is intended, (2) react adversely with the product, (3) malfunction for consumer use with regard to stability, and (4) to be non-machinable because of dimensional deviations.

    a.   Labels block together, i.e., cannot be separated easily.

    b.   Curled labels.

    c.   Labels fade during accelerated product storage tests.

    d.   Labels out-of-specification dimensionally so as not to fit the container.

    e.   Incorrect copy on labels.

2.   *Major Appearance Defect–1. 5 AQL*

A major appearance defect is one which makes the label unacceptable to the consumer because (1) it causes the container to be inaccurately identified or defined with regard to use and contents and (2) detracts seriously from aesthetic appearance.

    a.   Wrong color label material.

    b.   Wrong color printing.

    c.   Missing or blocked letters or words.

    d.   Poor registration.

3.   *Minor Functional Defects–4. 0 AQL*

Minor functional defect is a specification deviation for which a compensation in processing equipment can be made so as to handle the label at a regular production rate.

    a.   Labels slightly curled.

    b.   Labels slightly out-of-specification.

4. Minor Appearance Defect—10. 0 AQL

A minor appearance defect is one which detracts only slightly from the aesthetic appearance of the container and does not interfere with the copy covering identity and contents.

    a.   Printing slightly off-color.

    b.   Printing slightly off registration.

    c.   Faint printing.

    d.   Minor printing defects, slightly blocked letters.

5
STATISTICAL PROCEDURE

I. INTRODUCTION

Statistical tests used in the laboratory are the subject for entire books. There is no intent here to duplicate those endeavors. Certainly a good knowledge of statistics is essential for anyone who is involved in quality management.

Most of the statistical procedures used by chemists and engineers are basically grade school arithmetic functions. Some few are quite complex and require a working knowledge of advanced mathematics. We shall concentrate here on the simple procedures, and list in the bibliography some books for those who care to go further and delve deeper.

Statistics, in common with other disciplines, employs its own language. It will be helpful if you develop a working knowledge of several of the more commonly used statistical terms.

Let's begin by defining some of the more basic terms:

Precision - A measurement of the degree of agreement be-
             tween replicate analyses.

Replicate - Multiple runs of the same test on the same sample.

Accuracy - A measurement of the difference between one's
            results and the "true" answer.

Both precision and accuracy tell us how good our test is.
Precision also tells us how close the agreement will be among
separate analyses of the same sample. Accuracy tells us how
closely we can approach the true answer.

In dealing with the precision of an analysis, we must also be
concerned with repeatability and reproducibility.

Repeatability -   Refers to the precision of a single analyst
                    using the same equipment.

Reproducibility - Refers to the precision between a number
                    of analysts performing the same analysis
                    with different equipment.

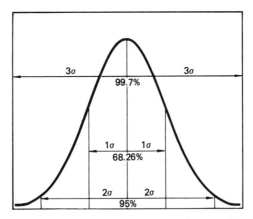

Figure 1. A standard distribution curve with 1- to 3- sigma limits.
(Reprinted by special permission from CHEMICAL ENGINEERING,
June 16, 1980, by McGraw-Hill, Inc., New York, N.Y. 10020.)

Early in any discussion of statistics one is generally con-
fronted with the term "normal distribution," followed by a drawing
of a bell-shaped curve. (Fig. 1)

What the curve shows is that if we run a number of tests,
using reasonably precise techniques, the distribution of the test
data will plot a bell shaped curve such as shown in Figure 1. The
bulk of the data will group around the average, while relatively
few results will be smaller or larger than the average.

If your precision is good, the curve will tend to be tall and
narrow. If your precision is poor, the curve will be short and
wide.

II. MEASUREMENT OF PRECISION

You will note in the Figure 1 curve that the test results are distributed with 1- to 3- sigma limits, shown in the curve as 1σ, 2σ
and 3σ. These refer to the standard deviation.

From earliest times chemists, and other scientists, had measured their data precision by means of the average deviation from
the overall average of a group of results. Plus and minus signs
were ignored in the calculation.

The procedure was simple, rapid and totally without mathematical basis. Don't use it. It can be grossly misleading.

The standard deviation (also called sigma) is the square root
of the sum of the squares of the differences between the individual
analyses and the overall average:

$$s = \sqrt{\frac{\sum (x^2)}{(n-1)}}$$

Where: s = standard deviation

x = deviations from arithmetic mean

n = total number of items

Variance, or dispersion, is the square of the standard deviation.

Referring back to Figure 1, note that if the data are normally distributed, a standard deviation of ± 1 (that is, one sigma) will include 68% of all values; ± 2 standard deviations (2 sigma) will include 95% and ± 3 standard deviations (3 sigma) will include 99.7%.

A word of caution: 1, 2 and 3 sigma values do not connote accuracy of a set of numbers; that is, a 2- sigma range does not mean 95% accuracy. It simply means that 95% of a set of data will fall within the arithmetic limits set by a range of two standard deviations.

Stated another way, if the distance equal to one standard deviation is measured off on the X axis of both sides of the arithmetic mean in a normal distribution, approximately 68% of the values will be included within the limits indicated.

If 2 standard deviations (2 sigma) or 3 standard deviations (3 sigma) are thus marked off, 95% and 99.7%, respectively, of all values will be included within those limits.

For most laboratory work, the 95% limit will suffice.

In practical terms, the typical sample represents a very small portion of the product produced by a process, or stored in a tank or shipped in a tank car or tank truck. It is, of course, impractical to analyze every milliliter or gram of every product manufactured or stored or shipped.

To compensate mathematically for the imbalance between sample size, and the total amount of product manufactured, stored or shipped, we can calculate standard deviation ("S") by using (n-1) in the division instead of (n).

III.   CONFIDENCE (PROBABILITY) LIMITS

It is useful to define your working range so that you can be confident 90%, 95% or 99% of the time that all of the values you obtain will fall within calculated limits. Said differently, 10%, 5% or 1% of the time, some values will not fall within calculated limits.

Caution: This is not to say that each result has a 90%, 95% or 99% chance of falling within calculated limits. Such a chance is only 50% because a result is either in or out of limits.

In calculating confidence or probability limits, there is a need to introduce a proportioning factor, necessary because, in general. only a few data out of a total population are being used.

This factor is derived from Table I, the Table of "t" probability level. The factor enables us to say that some percentage of

Table I. Table of probability level. (Reprinted
by special permission from CHEMICAL ENGIN-
EERING, June 16, 1980, by McGraw-Hill, Inc.,
New York, N.Y. 10020.)

Degrees of freedom	95%	98%	99%
1	12.71	31.82	63.66
2	4.30	6.96	9.92
3	3.18	4.54	5.84
4	2.78	3.75	4.60
5	2.57	3.36	4.03
6	2.45	3.14	3.71
7	2.36	3.00	3.50
8	2.31	2.90	3.36
9	2.26	2.82	3.25
10	2.23	2.76	3.17
11	2.20	2.72	3.11
12	2.18	2.68	3.06
13	2.16	2.65	3.01
14	2.14	2.62	2.98
15	2.13	2.60	2.95
16	2.12	2.58	2.92
17	2.11	2.57	2.90
18	2.10	2.55	2.88
19	2.09	2.54	2.86
20	2.09	2.53	2.84
21	2.08	2.52	2.83
22	2.07	2.51	2.82
23	2.07	2.50	2.81
24	2.06	2.49	2.80
25	2.06	2.48	2.79
26	2.06	2.48	2.78
27	2.05	2.47	2.77
28	2.05	2.47	2.76
29	2.04	2.46	2.76
30	2.04	2.46	2.75
35	2.03	2.44	2.72
40	2.02	2.42	2.71
50	2.01	2.40	2.68
60	2.00	2.39	2.66
70	2.00	2.38	2.65
80	1.99	2.38	2.64
90	1.99	2.37	2.63
100	1.98	2.36	2.63
200	1.97	2.35	2.60
500	1.96	2.33	2.59
1,000	1.96	2.33	2.58
∞	1.96	2.33	2.58

the time, values will fall within certain limits. These limits are
expressed as X% Confidence Limits.

In the "t" table, the first column indicates degrees of freedom which is equivalent to one less than the number of runs of a
particular laboratory test. The other columns represent the desired probability level.

To calculate the confidence limits, use the following formula:

$$C.L. = \bar{X} \pm \frac{ts}{\sqrt{N}}$$

$\bar{X}$ = arithmetic average of data

t = factor from Table I

s = standard deviation

n = number of tests run

Example:

Run No.	Data	Average $\bar{X}$	Deviations from Average	Deviations Squared
1	9		-1	1
2	12		+2	4
3	10		0	0
4	9		-1	1
	40/4 = 10	10	4/4 = 1	6

Average Deviation   = ± 1

Standard Deviation = $\sqrt{\dfrac{6}{4-1}}$ = ± 1.41

95% C.L. $= 10 \pm \dfrac{3.18 \times 1.41}{\sqrt{4}}$

95% C.L.   = 7.8 to 12.2

We can say that 95% of the time the data in the runs shown will fall within the limits 7.8 to 12.2.

IV.   COMPARISON OF ANALYSTS

It sometimes becomes necessary to compare two chemists who have analyzed the same material by the same method.

This can be done statistically by using the "t" test. Here is an example:

Assume m runs of Analyst A with average = $\overline{X}$; and n runs of Analyst B with average = $\overline{Y}$

Calculate, $t = \dfrac{\overline{X} - \overline{Y}}{S} \sqrt{\dfrac{mn}{m+n}}$

The formula used for calculating the standard deviation given earlier was based on a normal distribution. Since the distribution in this case is probably non-normal, we will use a somewhat different formula:

$$S = \left[\frac{\sum (X_1)^2 + \sum (X_2)^2}{(n-1) + (m-1)} \right]^{\frac{1}{2}}$$

Then, using "t" table (Table I) at the calculated D.F. line (degrees of freedom, which is one less than the number of analyses), find the given "t" value under the probability level that you have chosen. If this "t" value is larger than the one calculated from the above formula, there is virtually no probability of significant difference between the two analysts.

However, should the number be smaller than the calculated level, the two analysts probably differ.

Example:

	Analyst A			Analyst B	
Number of Analyses 7				5	

	d	d^2		d	d^2
5.90	0.10	0.0100	5.25	0.03	0.0009
5.97	0.17	0.0290	5.34	0.12	0.0144
5.75	-0.05	0.0025	5.11	-0.11	0.0121
5.73	-0.07	0.0049	5.18	-0.04	0.0016
5.74	-0.06	0.0036	5.22	0.00	0.0000
5.69	-0.11	0.0121			
5.85	0.05	0.0025			

5.80 Average   5.22 Average
     Sum of Squares 0.0646   Sum of Squares 0.0290

$$S^2 = \frac{0.0646 \times 0.0290}{6+4} = 0.0094 \text{ (variance)}$$

$$S = 0.0969 \qquad\qquad \text{(standard deviation)}$$

$$t = \frac{5.80 - 5.22}{0.0969} \sqrt{\frac{7 \times 5}{7+5}}$$

$$t = \frac{0.58 \times 1.709}{0.0969} = 10.23$$

Referring now to the "t" table at degrees of freedom = 11, under the 95% column we find that "t" = 2.20. This is obviously much smaller than the calculated 10.23, thus making apparent the very significant difference between the analysts.

V.   COMPARISON OF METHOD PRECISION

There will be occasions when it may be desirable to determine whether there is any difference in the precision of two analytical methods; or the possible effect of a modification in an analytical procedure. This can be determined by the use of the "f" test (named after R. A. Fischer, an English statistician), which is the determination of the variance of the methods.

You will recall that variance is equal to the square of the standard deviation:

Variance = S^2.

Table II.f distribution is useful for determining if there is a real difference between two analytical methods. (Reprinted by special permission from CHEMICAL ENGINEERING, June 16, 1980, by McGraw-Hill, Inc., New York, N.Y. 10020.)

Denominator df	\ Numerator df → 1	2	3	4	5	6	7	8	9	10	12	15	20	24	30	40	60	120	∞
1	161	200	216	225	230	234	237	239	241	242	244	246	248	249	250	251	252	253	254
2	18.5	19.0	19.2	19.2	19.3	19.3	19.4	19.4	19.4	19.4	19.4	19.4	19.4	19.5	19.5	19.5	19.5	19.5	19.5
3	10.1	9.55	9.28	9.12	9.01	8.94	8.89	8.85	8.81	8.79	8.74	8.70	8.66	8.64	8.62	8.59	8.57	8.55	8.53
4	7.71	6.94	6.59	6.39	6.26	6.16	6.09	6.04	6.00	5.96	5.91	5.86	5.80	5.77	5.75	5.72	5.69	5.66	5.63
5	6.61	5.79	5.41	5.19	5.05	4.95	4.88	4.82	4.77	4.74	4.68	4.62	4.56	4.53	4.50	4.46	4.43	4.40	4.37
6	5.99	5.14	4.76	4.53	4.39	4.28	4.21	4.15	4.10	4.06	4.00	3.94	3.87	3.84	3.81	3.77	3.74	3.70	3.67
7	5.59	4.74	4.35	4.12	3.97	3.87	3.79	3.73	3.68	3.64	3.57	3.51	3.44	3.41	3.38	3.34	3.30	3.27	3.23
8	5.32	4.46	4.07	3.84	3.69	3.58	3.50	3.44	3.39	3.35	3.28	3.22	3.15	3.12	3.08	3.04	3.01	2.97	2.93
9	5.12	4.26	3.86	3.63	3.48	3.37	3.29	3.23	3.18	3.14	3.07	3.01	2.94	2.90	2.86	2.83	2.79	2.75	2.71
10	4.96	4.10	3.71	3.48	3.33	3.22	3.14	3.07	3.02	2.98	2.91	2.85	2.77	2.74	2.70	2.66	2.62	2.58	2.54
11	4.84	3.98	3.59	3.36	3.20	3.09	3.01	2.95	2.90	2.85	2.79	2.72	2.65	2.61	2.57	2.53	2.49	2.45	2.40
12	4.75	3.89	3.49	3.26	3.11	3.00	2.91	2.85	2.80	2.75	2.69	2.62	2.54	2.51	2.47	2.43	2.38	2.34	2.30
13	4.67	3.81	3.41	3.18	3.03	2.92	2.83	2.77	2.71	2.67	2.60	2.53	2.46	2.42	2.38	2.34	2.30	2.25	2.21
14	4.60	3.74	3.34	3.11	2.96	2.85	2.76	2.70	2.65	2.60	2.53	2.46	2.39	2.35	2.31	2.27	2.22	2.18	2.13
15	4.54	3.68	3.29	3.06	2.90	2.79	2.71	2.64	2.59	2.54	2.48	2.40	2.33	2.29	2.25	2.20	2.16	2.11	2.07
16	4.49	3.63	3.24	3.01	2.85	2.74	2.66	2.59	2.54	2.49	2.42	2.35	2.28	2.24	2.19	2.15	2.11	2.06	2.01
17	4.45	3.59	3.20	2.96	2.81	2.70	2.61	2.55	2.49	2.45	2.38	2.31	2.23	2.19	2.15	2.10	2.06	2.01	1.96
18	4.41	3.55	3.16	2.93	2.77	2.66	2.58	2.51	2.46	2.41	2.34	2.27	2.19	2.15	2.11	2.06	2.02	1.97	1.92
19	4.38	3.52	3.13	2.90	2.74	2.63	2.54	2.48	2.42	2.38	2.31	2.23	2.16	2.11	2.07	2.03	1.98	1.93	1.88
20	4.35	3.49	3.10	2.87	2.71	2.60	2.51	2.45	2.39	2.35	2.28	2.20	2.12	2.08	2.04	1.99	1.95	1.90	1.84
21	4.32	3.47	3.07	2.84	2.68	2.57	2.49	2.42	2.37	2.32	2.25	2.18	2.10	2.05	2.01	1.96	1.92	1.87	1.81
22	4.30	3.44	3.05	2.82	2.66	2.55	2.46	2.40	2.34	2.30	2.23	2.15	2.07	2.03	1.98	1.94	1.89	1.84	1.78
23	4.28	3.42	3.03	2.80	2.64	2.53	2.44	2.37	2.32	2.27	2.20	2.13	2.05	2.01	1.96	1.91	1.86	1.81	1.76
24	4.26	3.40	3.01	2.78	2.62	2.51	2.42	2.36	2.30	2.25	2.18	2.11	2.03	1.98	1.94	1.89	1.84	1.79	1.73
25	4.24	3.39	2.99	2.76	2.60	2.49	2.40	2.34	2.28	2.24	2.16	2.09	2.01	1.96	1.92	1.87	1.82	1.77	1.71
30	4.17	3.32	2.92	2.69	2.53	2.42	2.33	2.27	2.21	2.16	2.09	2.01	1.93	1.89	1.84	1.79	1.74	1.68	1.62
40	4.08	3.23	2.84	2.61	2.45	2.34	2.25	2.18	2.12	2.08	2.00	1.92	1.84	1.79	1.74	1.69	1.64	1.58	1.51
60	4.00	3.15	2.76	2.53	2.37	2.25	2.17	2.10	2.04	1.99	1.92	1.84	1.75	1.70	1.65	1.59	1.53	1.47	1.39
120	3.92	3.07	2.68	2.45	2.29	2.18	2.09	2.02	1.96	1.91	1.83	1.75	1.66	1.61	1.55	1.50	1.43	1.35	1.25
∞	3.84	3.00	2.60	2.37	2.21	2.10	2.01	1.94	1.88	1.83	1.75	1.67	1.57	1.52	1.46	1.39	1.32	1.22	1.00

Degrees of freedom for numerator

Degrees of freedom for denominator

Interpolation should be performed using reciprocals of the degrees of freedom.

To use the "f" test:

1. Determine the standard deviation of the sets of data from each of the methods, or modifications of a method.

2. Determine the variance of each of the methods or modifications. Variance, you will recall, is the measure of dispersion or variability of data.

3. Divide the larger variance by the smaller.

4. Enter the f distribution table (Table II) at the desired level. The numerator degrees of freedom (n-1) are read across and the denominator degrees of freedom (m-1) are read down. At the intersection, read the f value.

   If this value is larger than the calculated ratio (Step 3 above), there is no apparent difference in the precision If the f value is smaller, there is a significant difference in the precision values between the two methods.

Example:

Assume two methods, call one "approved" and one "alternative." The following data are determined through eleven runs of both methods by the same analyst.

	Approved	Alternative	Difference
	1.8000	1.4000	0.4000
	1.6000	1.2000	0.4000
	1.4000	1.2000	0.2000
	1.4000	1.2000	0.2000
	4.2000	4.0000	0.2000
	4.0000	4.0000	0.0000
	4.0000	3.8000	0.2000
	4.2000	4.0000	0.2000
	4.2000	4.0000	0.2000
	1.4000	1.4000	0.0000
	1.6000	1.4000	0.2000
Sum:	29.8000	27.6000	2.2000
Mean:	2.7090	2.5090	0.2000
Std. Dev.:	1.2944	2.3276	
Variance:	1.6750	1.7626	
f Ratio:	1.0520 (calculated)		
F Table:	2.8200		

Since the "f" value from the table is larger than the calculated ratio, there is no apparent difference between the two
methods.

Sometimes it becomes important to determine whether certain
data obtained during process or quality control are valid. If not
valid you will very likely wish to discard them, or at least minimize their effect on your total data.

There is a simplified statistical device, sometimes called the
"Q" test which can help you determine which numbers you can
keep, and which you may, with some degree of safety, reject. [1,
2]

1. Calculate the difference of a doubtful number from the
    next closest number.

2. Divide this difference by the range (largest number
    minus the smallest).

3. If "Q" exceeds the values in the following table, the
    questionable number may be rejected with either 95% or
    99% confidence.

If, in a series of tests, we obtained the following data: 5,
14, 4, 8, 5, 3, 6, 4, we ask whether the largest number (14) may
be rejected.

Table III.

No. of Observations	Probability Level	
	95%	99%
3	.94	.99
4	.76	.89
5	.64	.78
6	.56	.70
7	.51	.64
8	.47	.60
9	.42	.57

$$Q = \frac{14 - 8}{14 - 3} = 0.55$$

for n = 8, Q at 95% = 0.47

Since the calculated value is larger, we may safely reject the value 14.
Applying the same technique to the value 8

$$Q = \frac{8 - 6}{8 - 3} = 0.40$$

from the table n=7, the Q value is 0.51. Since the calculated value is smaller, we cannot reject the number 8.

VI.   SIMPLE STANDARD DEVIATION TEST

Occasionally there may arise a need to estimate a "ball park" standard deviation without going through all the calculations.
The simplest procedure is to determine the range of results (smallest number subtracted from the largest number), and multiply by one of the following factors:

Table IV.

Range Estimate of Standard Deviation[1]

N	Factor
2	0.886
3	0.591
4	0.486
5	0.430
6	0.395
7	0.370
8	0.351
9	0.337
10	0.325

Example:

6 results are available: 4, 2, 5, 8, 6, 7

Estimated S = (8-4) x 0.395 = 1.6

S by calculation = 1.4

VII.   INTRODUCTION TO CONTROL CHARTS

The use of statistical quality control (SQC) to increase productivity through total control of quality is being much discussed
these days. There have been recent television programs and a
spate of articles in the daily press and periodicals devoted to the
incredible success of the Japanese in applying SQC principles to
their manufacturing operations.

In the months ahead, there will be increasing pressure by
American management to adapt SQC to American manufacturing
processes.

Most of the SQC used in Japan - and, for that matter, any-
where else in the world - has been in the area of hard goods manufacture, as opposed to the chemical process industries.

I think you will find, upon reflection, that it is much easier
to examine and control the quality of the raw materials, the process and the finished product in hard goods production, than in
the production of chemicals, either batchwise or continuous.

Yet, the need for improved quality and productivity in the
chemical process industries is evident.

In the beginning years of our space program, much publicity
was devoted to Zero Defects programs used in the manufacture of
spacecraft hardware. The need for quality control programs of
this intensity, for such products, is self-evident.

The applicability of such programs to the manufacture of
chemicals is somewhat less so. Chemicals, because of the nature
of the processes by which they are manufactured, will vary in
their properties. To account for this innate variation, a range of
acceptance is applied to the data for each property in a specification.

By charting the scatter of these data, we are able to chart
the variation in a process, and by controlling the variation, we
can control the quality of the product.

A.   Control Charts

The statistical concept involved in plotting these data is that of the control chart. Developed by W. A. Shewhart in 1931, it has found wide application in industry.

Variant techniques such as the Cumulative Sum Chart (Cum Sum), Acceptance Control Chart, Adaptive Control Chart and Geometric Moving-Average Chart have been well received and used. [3]

It is safe to say, in the chemical process industry, that variability is unavoidable in a manufacturing process - regardless of whether the process is continuous or batchwise. The variability can be caused by variation in raw materials, operations, humidity, process equipment - anything, in fact, that touches directly or indirectly upon the process.

The control chart is a graphical representation of these variations as they affect the quality of the product. (Fig. 2) It has several worthwhile advantages:

1.   Prediction of product quality trends; anticipating trouble before it occurs.

2.   Provides a sound basis for establishing or changing manufacturing specifications.

3.   Improving, narrowing, or otherwise changing min.-max. property limits.

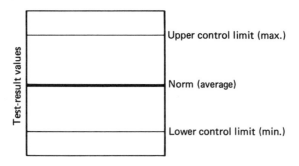

Figure 2. Shewhart quality-control chart. (Reprinted by special permission from CHEMICAL ENGINEERING, June 16, 1980, by McGraw-Hill, Inc., New York, N.Y. 10020.)

The basic control chart tells you whether variation is within previously determined upper and lower limits, and is thus acceptable, or outside these limits and thus cause for rejection and investigation into possible problems.

In our discussion on manufacturing specifications (Chapter I), we noted that the upper and lower control limits (or maximum and minimum) of a property in a manufacturing specification were set by the production department, with the concurrence, for analytical purposes, of the quality control laboratory.

When the process was new, the laboratory set up a blank chart (Fig. 2), analyzed a significant number of process samples for the property in question, and noted the data on the chart. If all the data fell between the upper and lower control limits, these limits were then considered realistic and acceptable.

If the data crowded the limit control lines, on one or both sides of the range, either the maximum and minimum specifications were unrealistic and in need of widening, or the process needed tighter control.

Should the majority of data fall outside one or both control limits, it could indicate that variables in the process need immediate attention.

The use of control charts is, of course, not restricted to the start of a new process.

They are valuable as checks for raw material or process changes; changes in operating conditions, or even changes in analytical procedures.

Data may be plotted as individual analysis results, as the average of individual results, or as the standard deviation of groups of those results. Where standard deviations are used, three sigma limits are the usual choice for control charts. Then 99.7% of the group averages would fall within the control limits, and only 0.3% outside. If two sigma limits were chosen, then 5% of the data would be outside the control limits. (Fig. 1)

Control charts should be kept as a permanent record of the quality testing of a product and attached to other pertinent product quality records. Notations may be made on the charts concerning causes of problems and their solutions for future reference.

Charts may be used to demonstrate to customers, active or potential, the past and present product quality history, and, as a result, possibly reduce the amount of sampling and property analysis which might be required by the purchaser.

VIII.   CALCULATORS AND COMPUTERS

Programmable calculators and desktop micro-computers are a factor in everyday business life. Lengthy mathematical relationships and laboriously repetitive analytical calculations can easily be performed by these devices.

These small instruments have the ability to store and analyze data for process variables or analytical control functions and return these data on an hourly, daily or, as required, on a long term basis.

Not only can numerical solutions be generated using these devices, but in many cases this output can be supplied in a form that can be included in a technical report. Printing and plotter attachments can display graphical analyses, frequently in multi-colors, using moving average charts, cumulative sum charts, adaptive control and acceptance control charts, all of which are widely used in data presentation.

It is often not necessary to develop programs for these desk top machines. Special purpose software programs are available in ever growing numbers. There are quality control packages, statistical analysis packages as well as software packages available with graphing capabilities to transform daily plant process data and laboratory analytical results into control charts (i.e. plotted graphs).

Control charts can also be generated by data captured, recorded, analyzed and displayed automatically through sensors located in or at pre-selected process points.

Obviously, where it is feasible to utilize them, computer developed control charts are faster and easier to use than manually developed charts. Though a whole segment of humor has developed concerning computer errors, control charts developed by graphic computers will be virtually error-free and will afford continuity and consistency.

The state of the art of computer technology and software design is constantly changing. What was difficult or unavailable yesterday is both simple and available today.

The costs of both hardware and software are coming down, and purchasing software programs for quality control and statistical analysis is easier than developing your own programs and less expensive as well.

A word of caution: In selecting a micro-computer (or programmable calculator) be sure you are buying a complete system, ready to use and easy to operate. Know what programs are immediately available to run on your system, whatever the language.

Don't count on developing your own programs (although you certainly may) or waiting for software programs to be developed. Assume that no more peripherals or optional accessories will be available for your machine than exist now.

Remember that no computer is any better than the data stored within it or fed to it. If you put in garbage, you get out garbage. And finally, be certain you really need a micro-computer before you buy one. Maybe a programmable calculator, though less glamorous (and a lot less expensive) will serve you as well. If you don't have reams of data to process, or have the need to develop only a few control charts per month, maybe you're better off doing the job manually.

REFERENCES

1. Dixon, W. J., "Processing Data for Outliers", Biometrics, March, 1953 pp. 74-80.

2. Dixon, W. J., Massey, F. J. Jr., "Introduction to Statistical Analysis", McGraw-Hill, New York 1957.

3. Freund, Richard A., "Control Charts Eliminate Disturbance Factors", Chemical Engineering, January 31, 1966 pp. 70-76.

SELECTED REFERENCES

Grant, E. L. and Leavenworth, R. S., "Statistical Quality Control", 5th ed., McGraw-Hill, New York, 1980.

Bader, M. E., "Quality Assurance - III, Using Statistics", Chemical Engineering, June 16, 1980 pp. 123-129.

Duncan, A. J., "Quality Control and Industrial Statistics", R. D. Irwin, Inc., Homewood, IL, 1953 pp. 420-421.

Feigenbaum, A. V., "Total Quality Control", McGraw-Hill, New York, 1961 pp. 275-277.

6
COMPLYING WITH
GOVERNMENT REGULATIONS

I. INTRODUCTION

It should come as no surprise to anyone in the chemical process industry that government — at any level — is a partner to the industry. This is particularly true in the areas of environmental affairs, health and consumer safety.

Over the past decade, the U.S. government has enacted a number of laws which regulate these areas. Beginning with the National Environmental Policy Act of 1969, we have had:

The Occupational Safety and Health Act	(1970)
The Clean Air Act	(1970)
Federal Water Pollution Control Act	(1972)
Consumer Product and Safety Act	(1974)
The Safe Water Drinking Act	(1974)

The Resource Conservation and Recovery Act   (1976)

The Federal Toxic Substances Control Act   (1976)

The Clean Water Act (amending the Federal   (1977)
Water Pollution Control Act)

This is not a complete list, of course, but it indicates those
pieces of legislation which are particularly pertinent to the chemical process industry.

Many of these regulations require that samples be taken and
chemically analyzed. In most plants, the control laboratory is the
only chemical laboratory at the site; thus, the work has fallen to
the lot of the quality assurance department. In many cases, its
personnel will also be assigned to taking the required samples.

Additionally, there may also be automatic monitoring instruments for air or water that will require attention - changing
charts, checking and calibrating.

The laboratory will be called upon to monitor a complex regulatory scene that, at times, will try patience, facilities and technology to the utmost.

For those who have not had wide exposure to the major environmental, health and consumer protection laws, a brief description of the major thrust of each may be helpful.

A.   National Environmental Policy Act (1969)

This Act, known as NEPA, derives its importance almost solely
from its environmental impact statement (EIS) requirements. All
federal agencies are required to prepare an EIS for every major
federal action significantly affecting the quality of the environment.

B.   Federal Water Pollution Control Act (1972)

The Act (FWPCA) provides that all discharges of pollutants into
navigable waters of the U.S. must meet prescribed effluent criteria. "Navigable waters' has since been defined by EPA as
virtually all surface water, navigable or not, in the U.S.

C. The Clean Water Act (1977)

This Act amended the FWPCA and authorized the EPA to set effluent standards and establish a national permit system. EPA has
set up the National Polutant Discharge Elimination System (NPDES)
providing for the issuance of permits for discharging effluents.
This is an area in which the laboratory is directly affected.
     I recommend that you obtain a copy of the toxic pollutant list
in 40 CFR 401. Sections 307 and 311 of the Clean Water Act list all
the chemical pollutants of which you should be aware.

D. The Clean Air Act (1970), amended 1977

This authorizes EPA to promulgate National Ambient Air Quality
Standards (NAAQS) to specify maximum concentrations of air
pollutants that are legally permissible. Pollutants covered include:
particulates, carbon monoxide, nitrogen oxide, sulfur dioxide,
lead, ozone, and hydrocarbons. Special standards have been set
for asbestos, benzene, beryllium, mercury and vinyl chloride.
This is another area in which the laboratory becomes directly involved.

E. The Safe Drinking Water Act (1974)

This was designed to protect drinking water systems, public and
private, as well as underground drinking water sources. The
drinking water in your plant is covered by this Act; as a result,
the laboratory is involved here, as well.

F. The Consumer Product Safety Act (1972)

This concerns the safety of products used by consumers in and
around their homes, or for recreational purposes. It includes
many chemical products: detergents, swimming pool chemicals,
fertilizers, brake fluids among others. Of course, the laboratory
is concerned with these.

G. Toxic Substances Control Act (1976)

Referred to as TSCA or TOSCA, and pronounced like the opera,
this act authorizes EPA to obtain production and test data from
the chemical industry on certain chemical substances, and to regulate these substances, as needed. Chemicals used exclusively in
food, pesticides, drugs and cosmetics are exempted from the Act,
but only becuase they are regulated by other federal legislation.

H. The Occupational Safety and Health Act (1970)

Known as OSHA, this Act is administered and enforced by the
Occupational Safety and Health Administration, under the Department of Labor. The Act deals with industrial hygiene and safety
in plants, for the protection of employees. For our purposes, we
shall discuss only the industrial hygiene aspects of the law, and
the role the plant laboratory may play in this area.
      These acts, and others, are facts of industrial life. It is not
likely they will go away. Their impact is felt by both large and
small companies. Compliance with the laws is an absolute necessity
and serves everyone well.
      Such compliance can involve workplace and breathing zone
air monitoring; as well as monitoring of drinking water and effluent discharges from the plant to sewers or streams.
      To provide assistance for such compliance, there has been
developed the:

I. National Institute for Occupational Safety and Health

Referred to, always, as NIOSH, this organization comes under the
Department of Health and Human Services (formerly HEW), Public
Health Service, Center for Disease Control. It is, in effect, the
technical development area of OSHA. It develops criteria for
recommending standards for occupational exposure to some 400
gases and vapors; and develops the sampling techniques and analytical methods for these substances.
      NIOSH has published a multi-volume "Manual of Analytical
Methods." These methods are necessary in evaluating the exposure of plant personnel to toxic substances. Without such methods,

it would be impracticable to determine compliance with OSHA permissible exposure limits.

Permissible exposure limits are referred to as "Threshold Limit Values (TLV)*. The TLV's refer to time weighted averages (TWA) of concentrations of airborne substances for an 8 hour day and 40 hour work week.

These figures are published in the Federal Register. In the event the Federal Register is not available to you, several organizations publish booklets or charts listing these values. [1]

MDA Scientific, Inc. [2] has a chart listing the OSHA TLV based upon the use of coconut shell charcoal fibers using the NIOSH recommended design.

Foxboro-Analytical [3] publishes a similar chart of time weighted average concentrations for over 300 substances, with analytical information based upon the use of the firm's Miran 1A, a general purpose gas analyzer.

There are several ways to sample air for toxic vapors. There are also several ways to analyze these samples once they are taken. These are described in detail in the NIOSH Manuals [4].

For sampling purposes, there are charcoal tubes (coconut fiber charcoal only), impinger tubes and detector tubes, which can collect the sample and analyze it colorimetrically. In these methods, a known volume of air is drawn through the tubes by small, battery-operated pumps.

The charcoal tubes, detector tubes, and pumps are available commercially from several sources. The impingers are available through glassware supply houses.

Analytical methods employ gas chromatographs, infrared spectroscopes, atomic absorption spectrophotometers and continuous paper tape monitors. Excepting the last, these methods are reasonably familiar and may be in use in your laboratory now.

Paper tape monitoring is different and, from the writer's experience, very good indeed. Essentially, it uses a chemically impregnated paper tape sensitized to and specific for a particular vapor or gas. The tape, supplied in a plastic cassette, moves at a constant rate and develops a colored stain proportional to the concentration of toxic vapor present in the air sampled. The monitor, in which the tape is designed to be used, employs a double beam principle in which the top half of the tape is exposed to the gas while the bottom half remains unexposed.

---

*TLV is a registered expression of the American Conference of Governmental Industrial Hygienists.

Detectors sense the light reflected from both portions of the
tape and produce a differential output signal. An advantage of
these paper tape monitors is that they are reasonably portable
within a plant. They may be battery operated and thus permit
readings to be made where electric current is not readily available,
as at the perimeter of plant property.

Monitors and tapes are available for a wide variety of chemicals at modest cost, and from several suppliers. We have found
the ones distributed by MDA Scientific, Inc. [2] to be very reliable.

II.   ANALYSIS OF WATER AND WASTES

In large measure, analytical procedures for water and wastes
(sewers, ditches, outfalls, potable water, etc.) are governed by
EPA recommendations [5]. Here EPA notes "instrumental methods
have been selected in preference to manual procedures because of
the improved speed, accuracy and precision."

In addition to these EPA procedures, there are two other
major approved sources of analytical techniques. Standard
Methods [6] and ASTM Methods [7]. In addition to instrumental
methods, the latter include standard, manually preferred chemical
procedures for laboratories not yet equipped for instrumental
analysis.

Present federal regulations, under Public Law 92-500, speci-
fically require the use of EPA approved methods of analysis.

Decision on plant or process modifications, construction of
new facilities, effectiveness of water treatment processes, among
others, may be based upon the results of analyses of water samples. The financial implications, alone, of such decisions dictate
the extreme care to be taken in analytical procedures.

EPA has published a handbook [8] that describes the basic
factors of water and wastewater measurement affecting the value
of the analytical results, and provides recommendations for the
control of the factors to ensure that analytical results are the best
possible. The handbook continues with a cogent observation:

"Regardless of which analytical methods are used in a
laboratory, the methodology should be carefully documented.
In some reports, it is stated that a standard method from
an authoritative reference was used . . . when close examination has indicated . . . that this was not true. Standard

methods may be modified or entirely replaced because of . . .
personal preferences of the laboratory staff. Documentation
of measurement procedures used in arriving at laboratory
data should be clear, honest and adequately referenced, and
the procedures should be applied exactly as documented."

An excellent rundown of instruments available for environmental monitoring is to be found in a 1979 Chemical Engineering
Deskbook on Instrumentation and Process Control [9].

If you don't already have them in your laboratory library, I
can't stress too highly, not only the desirability, but also the
necessity for owning all the manuals, charts, and handbooks
listed in the bibliography at the end of this chapter. The cost is
modest (some are free), the advantages and benefits of posses-
sion many.

In Chapter II, we noted the basic instrumentation commonly
used in quality control laboratories. Much the same instrumentation is basic to water and wastewater analysis.

This would include:

Analytical balance

pH meter

Conductivity meter

Turbidimeter

Spectrophotometer (infrared, atomic absorption)

Total carbon analyzer

Gas chromatograph

Selective ion electrodes

If you lack any of these instruments because they are not
necessary to your regular quality control effort, study your water
and wastewater analytical needs carefully — you may want to
acquire additional instrumentation.

A point in favor of instrumental analysis for those laboratories heavily involved in wet chemistry without instruments is
the fact that many of these devices — such as atomic absorption
spectrophotometers, selective ion electrode meters and Technician
Auto Analyzers can provide direct digital concentration data.

The data can be recorded directly on printout forms, without the need for additional calculation. This saves time and manpower and reduces error.

Where laboratory notebooks are concerned, these always should be bound books (not spiral or looseleaf), with sequentially numbered pages. All entries must be in ink, and each page should be signed by the analyst who performed the work.

A word of caution, at this point:

In environmental analysis, you are dealing, essentially, with very small concentrations of contaminants. In this area, parts per million is virtually macro-chemistry. You will be dealing in parts per billion and even parts per trillion.

In trace analysis, a trained, skilled analytical chemist is a necessity; as is the use of specific, approved methods and top-notch instrumentation. Don't undertake to provide environmental analytical data unless you are committed to make whatever effort is necessary to achieve reliable results.

Lacking such commitment, you'd be better advised to use the facilities of a commercial laboratory equipped to do the job.

A.   Sample Containers

EPA recommends that collection of water and wastewater samples be done with wide-mouth glass bottles equipped with screw caps containing Teflon liners. The use of Teflon liners is important since, without these, the sample could easily become contaminated by the adhesive used in holding the liner to the cap.

Polyethylene plastic bottles are not recommended because the water sample could leach trace quantities of plasticizer from them. Additionally, some contaminants could be absorbed on the poly-ethylene.

Glass bottles, of course, should be washed with chemical solutions to ensure freedom from organic contaminants. Caps and liners must be suitably washed and rinsed, as well.

To determine the effectiveness of your cleaning procedure, place distilled (or demineralized) water in a sample container for a period of time equal to that of a regular sample. Analyze this water for the contaminant in question and use this result as your blank.

B. Sampling

The objective of any sampling program involving water and waste-water is to obtain a representative sample of the environment in question. The basic sampling methods are grab sampling and composite sampling. Composite sampling may be performed manually or automatically. Grab samples are manually taken.

Manual sampling - grab or composite - should be considered when infrequent samples are required, when biological or sediment samples are needed, or where sites and conditions will not permit automatic samplers.

Automatic sampling is recommended when frequent samples are to be taken at a specific location. It is also recommended for outfall monitoring over a long period or where 24 hour composite samples are required.

This is not an either/or situation. Both sampling systems may be used at the same time in any given plant.

EPA leans toward the use of automatic samplers. They feel that "the use of automatic samplers eliminates errors caused by the human element in manual sampling, reduces personnel cost, provides more frequent sampling than practical for manual sampling, and eliminates the performance of routine tasks by personnel." [8]

While there are many automatic samplers available commercially, no one sampler is ideal for all situations. Select the automatic sampler most suitable for the water or wastewater to be characterized at your site.

In any sampling survey of water and wastewater, probably the single most important part is that of flow measurement. Without accurate flow measurement data, it will be impossible to certify on your NPDES report how much of which contaminant is leaving your plant in any given time period.

EPA again: "Flow measurement data may be instantaneous or continuous. For continuous measurement, a typical system consists of primary devices such as weirs and flumes and secondary devices such as flow sensors, transmitting equipment, recorders and totalizers. The improper installation or design of a primary device or malfunction of any part of a secondary device results in erroneous flow data. The accuracy of flow measurement data also varies widely, depending principally on the accuracy of the primary device and the particular flow measurement method used. In any case, an experienced investigator should be able to measure flow rates within ± 10 percent of the true values." [8]

C.   Recordkeeping

Authorized representatives of EPA are empowered to enter plant premises where effluent sources are located. They may inspect monitoring equipment and methods and may sample effluents.

An important note: if samples are taken by EPA, request that the samples be split so that you may run your own analysis.

The EPA representative may also inspect and copy your records on environmental data.

Probably the most important of these records, from the laboratory's point of view is the NPDES (National Pollutant Discharge Elimination System) permit. The data listed in this record are supplied by the laboratory to plant management and are based upon the analytical data produced on wastewater samples. These reports are filed quarterly by plant management.

It is essential that:

1.   the data in the NPDES report coincide with laboratory data,

2.   it be possible to trace report data back to the laboratory notebooks quickly and easily, and

3.   the data in the notebooks be accurate, clear, concise, pertinent and easily understood.

D.   Environmental Audits

Such an audit is especially advisable on at least a semi-annual basis in multi-plant, multi-laboratory situations. In a simple unit, the laboratory manager should be doing this on a continuing basis.

1.   Who took the sample, when was it taken, where was it taken and how was it taken?

2.   Does this conform with EPA, state, county or municipal requirements?

3.   Who took the flow measurement data? How was it taken, when was it taken, where was it taken?

4.   Does timing of flow data coincide with sampling time? If not, has EPA approved variance in permit?

5. Are laboratory data entered in ink in hardbound, sequentially numbered page notebooks used exclusively for this purpose?

6. Who did the calculations for reporting purposes?

7. Are the calculations shown in the notebook?

8. Have the calculations been checked by several persons before entry in report?

9. Have all samples been reported?

10. Does NPDES data compare with laboratory data?

III. GOOD MANUFACTURING PRACTICE

Title 21 of the Code of Federal Regulations (CFR) Parts 110 and 210-211 describes the general guidelines to be used in the manufacture of food additive chemicals and finished pharmaceuticals. In the latter instance, the Food, Drug and Cosmetic Act does not include drug intermediates within the definition of the term "drug". As a result, FDA's current good manufacturing practice regulations, and the current good manufacturing practice guidelines for finished pharmaceuticals do not apply to drug intermediate manufacture.

In the interest of product quality and minimizing the exposure to adulteration of a finished pharmaceutical product, the following general guidelines should be used when manufacturing, processing or packaging a drug intermediate.

A. Organization and Personnel

The responsibilities of the quality control unit are as follows: (a) The quality control unit shall have the responsibility and authority to approve or reject all components, product containers, closures, in-process materials, packaging materials and labelling. It shall have the authority to review production records to assure that no errors have occurred, or if they have, that they have been fully investigated. (b) Adequate laboratory facilities for product testing shall be available to the quality control unit. (c) Quality control shall have the responsibility for approving or rejecting all procedures or specifications pertinent to the identity, strength, quality and purity of the drug components. (d) Quality control's responsibilities and procedures shall be in writing.

B. Personnel Qualifications

Personnel responsible for direction, manufacture and quality control shall (1) be adequate in number and background of education, training and experience to assure quality of the product; (2) have capabilities commensurate with their assigned functions, understanding of the manufacturing in control operations they perform; and (3) have adequate information concerning the application of these practices to their function.

C. Personnel Responsibilities

(a) Personnel shall wear clean clothing appropriate for the duties they perform. (b) Personnel shall practice good sanitation and health habits. (c) Any person shown by medical examination or supervisory observation to have an apparent illness or open lesions that may affect safety or quality of the product shall be excluded from direct contact with the product until the condition is corrected.

D. Buildings and Facilities

(a) Buildings shall be maintained in a clean and orderly manner and be of suitable size, construction and location to facilitate adequate cleaning, maintenance, and operation. (b) Adequate space shall be provided for the orderly placement of equipment and materials to minimize risk of mix-ups between different product components, in-process materials, packing materials, or labeling, and to minimize the possibility of contamination. (c) Adequate lighting shall be provided in all areas. (d) Adequate ventilation shall be provided to (1) minimize contamination of products by extraneous adulterants, including cross-contamination by dust or particles; (2) minimize dissemination of microorganisms from one area to another; and (3) provide suitable storage conditions for components in-process materials, and finished materials. (e) Sewage, trash, and other refuse in and from the building and immediate premises shall be disposed of in a safe and sanitary manner. (f) Any building used in the manufacture, processing, packing or holding of a drug intermediate or component shall be maintained in a clean and sanitary condition and free of infestation by rodents, birds, insects and other vermin.

E. Equipment

(a) The equipment shall be so constructed that all surfaces that come into contact with the product shall not be reactive, additive, or absorptive so as to alter the safety, identity, strength, quality or purity of the drug components or intermediates beyond the official or other established requirements. Lubricants or coolants, etc., must not come into contact with components, containers, closures or in-process materials so as to alter the safety, identity, strength, quality or purity of the product. (b) Written procedures shall be followed for cleaning and maintaining equipment. (c) Automatic, mechanical and electrical equipment shall be routinely calibrated, inspected and checked according to a written program designed to assure proper performance. (d) Filtration, as needed, shall be through non-fiber releasing filters, unless it is not possible to manufacture the product without the use of such filters.

If the use of a fiber-releasing filter is required, an additional non-fiber releasing filter of maximum pore size of 0.22 microns (0.45 microns if the manufacturing conditions so dictate) shall subsequently be used to reduce the content of any asbestos-form particles in the drug intermediate or component.

F. Control of Components and Containers

(a) Written procedures shall be followed for the receipt, identification, storage, handling, testing and approval or rejection of components and containers. (b) Components, containers and closures must be handled and stored in a manner to avoid contamination at all times. This means storage off the floor and adequately spaced to permit cleaning and inspection. (c) Each lot of a component, container, or closure that is liable to contamination with

filth, insect infestation, or other extraneous adulterant shall be
examined against established specifications for such contamination.
(d) Rejected items should be identified and separated to prevent
their use in manufacturing or processing operations for which they
are unsuitable. (e) Containers and closures shall not be reactive,
additive, or absorptive so as to alter the safety, identity,
strength, quality, or purity of the product beyond the official or
established requirements.

IV. PRODUCTION AND PROCESS CONTROLS

Production and control procedures shall include all reasonable
precautions to assure that the product has the safety, identity,
strength, quality and purity it is purported to possess:
Each significant step in the process, such as the selection,
weighing, and measuring of components, the addition of ingredients during the process, weighing and measuring during various
stages of the processing, shall be performed by a competent and
responsible individual, or if such steps in the processing are controlled by precision automatic, mechanical, or electronic equipment,
their proper performance is adequately checked by one or more
competent and responsible individuals. The written record of the
significant steps in the process shall be identified by the individual charged with checking these steps. Such identifications shall
be recorded immediately following the completion of such steps.
To assure the uniformity and integrity of products, written
procedures shall be established and followed that describe adequate in-process controls. In-process sampling shall be done at
appropriate intervals using suitable equipment.
Procedures shall be instituted whereby review and approval
of all production and control records, including packaging and
labeling, shall be made prior to the release or distribution of a
batch.
Rejected in-process materials shall be identified and separated
to prevent their use in manufacturing or processing operations for
which they are unsuitable.
Written procedures shall be established and followed pre-
scribing a system for reprocessing batches that do not conform to
standards or specifications and the steps to be taken to insure
that the reprocessed batches will conform with all established
standards, specifications, and characteristics. No reprocessing
shall be performed unless approved by quality control.

V. PACKAGING AND LABELING CONTROL

Labeling controls shall include: 1. the maintenance and storage
of each type of label and package labeling representing different
products, strength, or quantity of contents in such a manner as
to prevent mix-ups and provide proper identification; and 2. a
suitable system for assuring that only current labels and packaging labeling are retained and that stocks of obsolete labels and
package labeling are destroyed.
    Procedures shall be established and followed to identify the
finished product with a lot or control number that permits determination of the history of the manufacture and control of the
batch. A year, day, hour or shift code is appropriate as a lot or
control number for drug intermediates manufactured or processed
in continuous product equipment.

VI. HOLDING AND DISTRIBUTION

Procedures shall be followed for the quarantine of product prior
to release by quality control and for the storage under appropriate
conditions of temperature, humidity and light.

VII. LABORATORY CONTROLS: RECORDS AND REPORTS

All specifications, standards, sampling plans, test procedures, or
other laboratory control mechanisms, including any changes to
specifications, etc., shall be drafted by the appropriate organiza-
tional unit and reviewed and approved by quality control. The
requirements in this subpart shall be followed and shall be documented at the time of performance. Any deviation from the written
specifications, standards, sampling plans, test procedures, or
other laboratory control mechanisms shall be recorded and justified.
    Laboratory controls shall include the establishment of scienti-
fically sound and appropriate specifications, standards, test procedures, and master records to assure that components, in-process and finished products conform to appropriate standards of
identity, strength, quality and purity.

1. For each batch of product there shall be appropriate
    laboratory determination of satisfactory conformance to
    final specifications, including the identity and strength
    of each active ingredient.

2. All sampling and testing plans shall be described in
    written procedures that include the method of sampling
    and the number of units per batch to be tested.

3. Products failing to meet established standards or specifi-
    cations and any other relevant quality control criteria
    shall be rejected. Reprocessing may be performed.

4. Master production quality control, distribution and com-
    plaint records should be retained and be readily available
    for authorized inspection during the appropriate reten-
    tion period.

VIII. GUIDELINES FOR PREPARATION OF FOOD CHEMICAL ADDITIVES

The following general guidelines for the preparation of food chemical additives have, similarly, been developed from the FDA's current good manufacturing practice regulations. They are, in some
ways, slightly less stringent than the GMP for drug intermediates.

A. Personnel

Plant management shall be responsible for assuring the following:

1. Disease Control

No person affected by disease in a communicable form, or while
the causes of such disease, or while affected with boils, sores, in-
fected wounds or other abnormal sources of bacterial contamination,
shall work in a chemical plant in any capacity in which there is a
reasonable possibility of ingredients or products becoming contaminated by such person, or of disease being transmitted to
other individuals.

2.   Cleanliness

Where direct contact with food chemicals is required, those persons in such contact shall take appropriate precautions to prevent contamination of chemicals with microorganisms or foreign substances which may affect the purity, safety, or otherwise adversely affect the fitness of such substance for food additive use.

B.   Buildings and Facilities

1.   Grounds

The location and ground surrounding a food additive plant shall be free from conditions incompatible with manufacturing, processing, packing or holding operations of food additives.

2.   Plant Construction and Design

Plant buildings and structures shall be of suitable size, construction and design to facilitate maintenance and operation for their intended purpose. The plant facilities shall: (a) provide sufficient space for orderly placement of equipment and storage of materials used in any of these operations; (b) provide adequate separation by partitions or by location so as to separate those operations which may cause cross-contamination of chemicals; (c) provide adequate lighting to all areas where chemicals are processed, examined or stored or where equipment and utensils are washed. Light bulbs, fixtures, skylights or other glass suspended over exposed chemicals in any step of preparation shall be of the safety type or otherwise protected to prevent chemical contamination in case of breakage; and (d) provide adequate ventilation and storage facilities to minimize objectionable odors and noxious fumes and vapors (including steam), in areas where they may contaminate chemicals.

C.   Sanitary Facilities and Controls

Each plant shall be equipped with adequate sanitary facilities and accommodations. Any water that contacts food chemicals shall be safe and of adequate sanitary quality.

D. Sanitary Operations

1. General Maintenance

Buildings, fixtures and other physical facilities of the plant shall
be maintained in such condition and cleaned in such a manner as
to allow routine production of food additives with minimal danger
of contamination. Detergents and disinfectants and other supplies
employed in cleaning and sanitizing procedures shall be safe and
effective under conditions of use. Any toxic chemicals used as
raw materials or intermediates or required for maintenance of sanitary conditions shall be utilized in such manner as to prevent contamination of finished product and/or the creation of a health hazard to employees or the public.

E. Equipment and Procedures

All plant equipment shall be suitable for its intended use, durable,
and kept in good repair. The design, construction and use of
such equipment shall preclude the adulteration of food additives
with lubricants, fuel, metal fragments, contaminated water or any
other contaminants.

Use of Polychlorinated Biphenyls. The following special provisions are necessary to preclude accidental PCB contamination in
the production, handling and storage of food additive products:

  a. New equipment and machinery for handling or processing
      food chemicals shall not contain PCB's.

  b. Existing equipment or machinery must be refilled with a
      heat exchange fluid that does not contain PCB's.

  c. Removal of any other PCB-containing materials wherever
      there is a reasonable expectation that such materials
      could cause food chemicals to become contaminated with
      PCB's either as a result of normal use or as a result of
      accident, breakage, or other mishap.

  d. The toxicity and other characteristics of fluids selected
      as PCB replacements must be adequately determined so
      that the least potentially hazardous replacement is used.

The foregoing provisions do not apply to electrical trans-
formers and condensers containing PCB's in sealed containers.

F.   Processes and Controls

All operations in the manufacturing, packaging and storing of
chemicals shall be conducted in accord with adequate sanitation
principles and under adequate, responsible supervision. All reasonable precautions, including the following, shall be taken to
assure that production procedures do not contribute to contamination of the processed product.

Raw materials and intermediates shall be inspected as necessary to assure that they are satisfactory for production of chemicals and shall be stored under conditions that protect against contamination and minimize deterioration.

Containers and carriers of raw materials shall be inspected
on receipt to assure that their condition has not contributed to the
contamination or deterioration of the products.

All chemical processing, including packaging and storage,
shall be conducted under such conditions and controls as are necessary to minimize the potential for contamination with chemicals and
other extraneous materials.

Recordkeeping and product coding shall be utilized to enable
positive lab identification or facilitate, where necessary, the segregation of specific lots that may have become contaminated or otherwise unfit for human consumption. Records shall be retained for
two years.

Storage and transport of finished products shall be under
such conditions as will preclude all contamination and will protect
against deterioration of the product and the container.

It can be seen, easily, that the provisions of the foregoing
Good Manufacturing Practice guidelines can be generally applicable
to the manufacture of virtually any chemical. Adherence to these
guidelines can only positively affect the quality and fitness for
use of your plant's products, even if they are not drug intermediates or food additives.

REFERENCES

1.   "Threshold Limit Values of Chemical Substances in Workroom

Air", American Conference of Governmental Industrial Hygienists, P.O. Box 1937, Cincinnati, OH 45201

2.   MDA Scientific, Inc., 1815 Elmdale Avenue, Glenview, IL
     60025.

3.   Foxboro Analytical, 140 Water St., P.O. Box 449, South
     Norwalk, CT 06856.

4.   National Institute for Occupational Safety and Health, "Manual
     of Analytical Methods", 2nd ed., Vols. 1-6, U.S. Dept. of
     Health and Human Services, Public Health Service, Center
     for Disease Control, National Institute of Occupational Safety
     and Health, Cincinnati, OH.

5.   "Methods for Chemical Analysis of Water and Wastes", Environ-
     mental Monitoring and Support Laboratory, Office of Research
     and Development, U.S. Environmental Protection Agency, EPA-
     600/4-79-020, Cincinnati, OH 45268, March, 1979.

6.   "Standard Methods for Examination of Water and Wastewater",
     14th ed., American Public Health Assn., 1015 18th St., N.W.,
     Washington, D.C. 20036.

7.   "Annual Book of Standards", Part 31: Water, American Ser-
     vice for Testing and Materials, 1916 Race Street, Philadelphia,
     PA 19103.

8.   "Handbook for Analytical Quality Control in Water and Waste-
     water Laboratories", EPA 600/4-79-019, Environmental Monitor-
     ing and Support Laboratory, U.S.E.P.A., Office of Research
     and Development, Cincinnati, OH 45268.

7
QUALITY COSTS AND QUALITY AUDITS

I. INTRODUCTION

The amount of investment required to achieve savings through reduction of quality costs is far less than that required for the same
return through increased sales or new product development.

It is entirely possible that most chemical processors have only
vague ideas concerning their quality costs. Many may consider
their quality control laboratory costs as the entire sum.

Obtaining quality cost estimates is often a difficult task since
many companies do not keep adequate records to enable proper
assessment of their quality losses. Some quality losses may be
charged to the cost of sales or transportation or diluted in production charges.

Since identifying the true sources of quality losses is a prime
requisite for cost reduction of your quality effort, it is necessary
to develop a reporting system which will identify losses and reveal
their responsibility in terms of departments, process equipment,
operators, raw material, product or any other logical subdivision.

Quality costs are probably dispersed in your company's various cost centers, departmental budgets and accounting system.

They begin back in the R&D laboratory and carry forward into pilot plant testing. They surface in manufacturing in the form of product which must be dumped, reworked or downgraded for sale.

In sales, quality costs are represented by the handling of customer complaints, as well as the sale, at discount, of nonspecification product. Transportation costs will include return shipments of unwanted product and possible additional costs involved in overpacking or repacking of opened or damaged containers.

Finally, there is the cost of the quality control laboratory itself; its staff, instrumentation and all the necessary items which enable it to function.

We can examine where these costs originate and how they may be controlled or minimized. There are inherent in what has been said above, three categories into which quality costs fall.

II.   PREVENTION, APPRAISAL AND FAILURE

The smallest category is, probably, prevention costs. These are primarily for the purpose of keeping defective products from being manufactured.

Under Prevention, we might have such items as:
1. Quality training for employees.
2. Cost of process equipment changes or improvements to reduce or prevent quality problems.
3. Cost of product above that which is required by your label and is supplied to the customer without extra charge because of less than perfect process or analytical control.

Appraisal and Failure (both internal and external) are obviously much larger quality cost segments.

Under appraisal costs might be the following:
1. Cost involved in raw material sampling and examination.
2. Cost involved in process control sampling and analysis.
3. Cost involved in finished product sampling and analysis.

4. Cost of the quality control laboratory; staff, equipment, instrumentation, etc.
5. Cost of maintaining and calibrating instruments.
6. Cost of maintaining records and data; as well as retention of production or shipping samples.
7. Fees for outside or referee laboratory use.
8. Costs involved in meetings with customer quality, purchasing or marketing representatives on matters pertaining to product specifications or quality of product sold to them.

Without distinguishing between internal and external Failure, these might be the cost of reworking a nonspecification product making it salable; the cost of dumping, or otherwise disposing of, a product that cannot be reworked. (This must include cost of raw materials, processing and handling, as well as the actual disposal cost of the material to be dumped.); cost of internal meetings, writing of reports, explaining why "it" happened, and possible visits to raw material vendors looking for somebody to blame for the problem; money spent to placate unhappy customers, whether or not the expense is justified (cost of visits, investigation of the problem, writing reports, and thinking of ways to avoid the blame); the financial loss incurred by selling an off-specification product instead of specification material; and costs involved in analysis of customer complaints.

The items noted above are not intended to be all inclusive. You can almost certainly think of other costs.

A word of caution: quality costs should not necessarily be considered as absolute numbers. Some are not always easy to quantify, some may be diluted or buried deeply within other cost centers. Take those costs which are solid and easy to obtain month after month and examine their trends. Are they increasing, decreasing, remaining steady? Which appear to be your major problems, recurring often; which are minor or trivial, or occur infrequently. Don't allow yourself to become bogged down in the examination of many small complaints. The cost of such examination can exceed the quality loss. Concentrate instead on the relatively few major problems.

The aggregate of these quality costs has been variously estimated as being as high as 50% of manufacturing costs. They may well be almost certainly in the 5-15% range. Even in a small company, this can translate into significant dollars.

III. COST REDUCTION PROGRAM

Any worthwhile quality cost reduction program must have as its
nucleus a competent, adequately staffed quality control department.
Depending on the size of your company, its product mix and sales,
such a staff can be two people or twenty.

Whatever the size, the staff must be trained, competent and
professional. In all too many companies, quality control programs
have been led by poorly qualified technicians reporting to supervisors who, themselves, have no knowledge of, interest in, or
commitment to quality. This is obviously counterproductive.

To begin with, your program should have a direction or goal.
The goal should be significant quality cost reductions, possibly of
the order of 25, 50 or even 75%.

Having identified the sources of your quality cost losses, try
to attack only the major ones. Don't spin your wheels and address
total effort against every problem and every loss. Apply your
energy to the high loss areas.

Determine the cause of your problem. Is it raw material; a
process aberration; equipment oriented; untrained operators?
Whatever it is, determine that it is not a one time quality problem;
such as a temporary process upset, or a new operator filling in at
an unfamiliar job.

Determine the action to be taken. Don't waste your time in
prolonged studies of the problem in committee meetings. Keep
these to an absolute minimum. They only involve time and increase
the dollar ton total. Quality control may aid in pointing out the
problem area, but is not responsible for correcting it. Plan your
action, make assignments and then see to it that assignment sched-
ules are followed.

The effort may involve installation of new equipment, changing
a process, or retraining operators to a new system. Once the
problem has been solved, and your quality loss minimized, don't
continue to attack the problem. Withdraw some of your forces,
decrease the concentration of your effort. Don't continue to run
endless samples through the laboratory. This costs money, too.

To assure that the problem remains in control, institute a
follow-up audit system. Check your customer complaints regularly.
Continue sampling and analyzing raw materials and finished goods.
Watch your control charts for signs of slippage.

Lest there be any thought, because of the above, that management is not involved in the causes of quality losses, let it be said
that it is a management responsibility to provide plant personnel
with a workable process and the best equipment designed to
operate this process and produce acceptable quality material.

IV.   WEIGHTS AND MEASURES

One of the functions of the quality control department of your
plant is to check on the fill of containers. It serves no useful
purpose to manufacture acceptable product, then alienate a
customer by short-filling the containers shipped to him. Converse-
ly, the situation is not improved, from your point of view, by over-
filling and "giving away the store."

Actually, a prearranged small overfill can be desirable. Basically, this is done to prevent — or, at least, minimize — under-
filling.

The National Bureau of Standards publishes a Handbook (#133)
"Checking the Net Contents of Packaged Goods" which is used as
a guide for federal and state inspectors whose function it is to
check on package weights of finished goods. This was published
in June 1981 and replaces Handbook #67. A copy may be obtained
through the Superintendent of Documents, U.S. Government
Printing Office, Washington, D.C. 20402.

The handbook is valuable, though unnecessarily complicated
and somewhat difficult to follow; but, if you use the tables for
maximum allowable weight variations for packaged goods you can
save money and keep out of trouble.

The control laboratory at your plant should spot-check the
weights of finished packages, either as they come off the production line or in the warehouse prior to shipment. Additionally,
the laboratory should conduct a routine periodic audit of all the
scales used in the plant to weigh either raw materials or finished
product. Figure 1 is a form that can be used for such an audit.
In Section I, we list each size, by name and serial number and
give other information about it. In Section II, a random number
of containers, listed by product, are weighed and the actual
weight compared to the label (claimed) weight. Section III can be
used for comment, or to describe the weighing procedure used.

Such an audit should be performed as frequently as possible.
Most plants have an outside organization come in and clean and
calibrate the scales periodically. But since this is usually done
only once or twice a year, it is important to check between times.

The National Bureau of Standards publishes NBS Handbook
#44 (1982), "Specifications, Tolerances, and Other Technical Re-
quirements for Weighing and Measuring Devices." It is available
through the Superintendent of Documents, U.S. Government
Printing Office, Washington, D.C. 20402.

This one is valuable and highly recommended, and is especially worthwhile for the tolerance tables for scales and the general
tables of weights and measures.

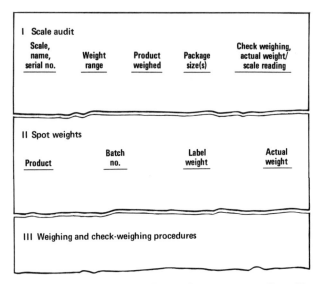

Figure 1. Quality-assurance weights and measures audit. (Reprinted by special permission from CHEMICAL ENGINEERING, June 16, 1980, by McGraw-Hill, Inc., New York, N.Y. 10020.)

V. QUALITY AUDITS

Such an audit or survey is an appraisal of a plant's (or company's) system of controlling the quality of its output. This audit is particularly valuable in a multi-plant company where a headquarters staff is responsible for the quality of products manufactured in all the plants. But it also has much value in even a single production unit.

An audit should be conducted, at the very least, on an annual basis. More complex plants may require a full or partial audit somewhat more frequently.

The audit must cover every facet of quality control, including those functions or departments that have only a peripheral impact upon quality. And nowadays, when the control laboratory is called upon to analyze effluent streams to provide data for environmental reports submitted to governmental agencies, this function must be audited as well.

What follows is a set of guidelines that can be used as a basis for what is to be sought in a quality audit. Obviously, depending upon time and circumstances, all, or any sections of this audit may be performed.

VI.   AUDIT GUIDELINES

A.   Raw Material Specifications

    a.   Is there a specification for each product?

    b.   Is there an analytical method for each of the critical parameters?

    c.   Who samples the material?

    d.   When, how and where is the sample taken?

    e.   Is the specification realistic?

    f.   How do analytical data obtained by your laboratory compare with specifications?

    g.   What is the date of the most recent specification review?

    h.   Does the purchasing department have a copy of each specification?

B.   Manufacturing Specifications

    a.   Is there a specification for each product?

    b.   Is there an analytical method for each of the critical parameters?

    c.   Who samples the product? Lab, production, shipping?

    d.   When, how and where is the sample taken? How often?

    e.   Is the specification realistic?

    f.   Check random production analytical data over recent months and note conformance to, or variance from, specification.

    g.   What is the date of the most recent specification review? (It should be reviewed and updated every 4-5 years.)

C.   Product Identification

    a.   Are all products completely coded: by day, week, month, year, shift, plant?

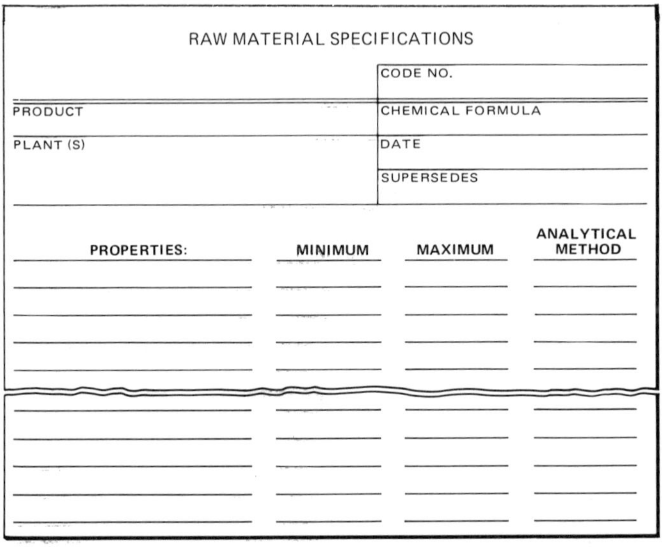

RAW MATERIAL SPECIFICATIONS

CODE NO.

PRODUCT

CHEMICAL FORMULA

PLANT (S)

DATE

SUPERSEDES

PROPERTIES:	MINIMUM	MAXIMUM	ANALYTICAL METHOD

Figure 2. Sample forms for use in setting up specifications for manu-factured products and for raw materials. (Reprinted by special permis-sion from CHEMICAL ENGINEERING, February 11, 1980, by McGraw-Hill, Inc., New York, N.Y. 10020.)

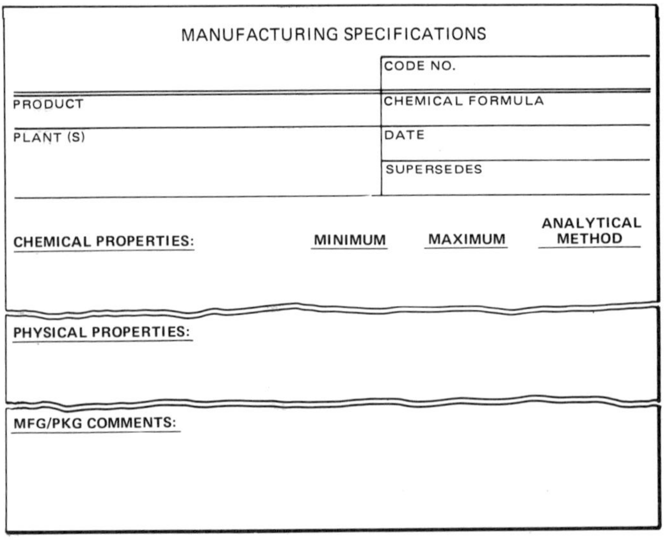

MANUFACTURING SPECIFICATIONS

CODE NO.

PRODUCT

CHEMICAL FORMULA

PLANT (S)

DATE

SUPERSEDES

CHEMICAL PROPERTIES:	MINIMUM	MAXIMUM	ANALYTICAL METHOD

PHYSICAL PROPERTIES:

MFG/PKG COMMENTS:

Figure 3. (Reprinted by special permission from CHEMICAL ENGINEERING, February 11, 1980, by McGraw-Hill, Inc., New York, N.Y. 10020.)

    b. If not, what is the coding status?

    c. Are codes legible, indelible, easily available on container?

    d. How are bulk shipments coded? Shipments by cars, trucks, pipeline, etc?

    e. What precautions does the plant use to prevent uncoded containers from being shipped?

    f. Can product be traced back to retained sample?

D. Sample Retention

    a. What production unit do samples represent? Shift, batch, lot, composite?

    b. Are samples identified properly? Labels, code, date?

    c. Are they in proper and adequate containers suitable for required storage period?

    d. Are they readily available in storage?

    e. Is storage facility (area) adequate? Is electricity, water, and adequate, safe and approved drainage available. Are there adequate shelves, and ventilation?

E. Data Retention

    a. Are laboratory control data retained? Where? For how long? In what form?

    b. Is notebook format satisfactory for data retrieval? Are data clearly presented?

    c. Are notations in ink? Pages numbered? No erasures?

    d. Are notebooks signed on each page by chemist, analyst or technician?

F. Shipping Documents

    a. Are all shipments completely identified with product batch or lot number, or other system of identification?

    b.   Is number of containers of each batch or lot number noted for each shipment? (This can be important in the event of a recall.)

G. Control Laboratory

    a.   Have analytical methods been checked for accuracy and precision? How? By whom?

    b.   Have chemical standards solutions been checked? How often? By whom?

    c.   In multi-plant companies, are methods for similar products the same in all laboratories?

    d.   Who samples production initially? Are they trained and competent?

    e.   Who takes resamples? (If lab doesn't take initial sample, quality control personnel should be involved in resampling.)

    f.   If product sample is out of specification, how many resamples are permitted to be taken before a product is finally accepted or rejected? (Only one resample per batch or lot should be permitted.)

    g.   Is the sampling method appropriate for the product?

    h.   Are statistical methods used to define product quality and/or report results?

    i.   Does laboratory check the weight of finished product containers prior to shipment, or while in inventory?

    j.   Has a weight and measures audit been performed on plant scales within past ninety days?

    k.   Does the laboratory have the analytical capability to control the products manufactured in its plant?

H. Shipping Containers

    a.   Does the plant use check sheets to note condition of cars and trucks (or barges) prior to loading?

b. Are the cars and trucks (or barges) actually inspected?

c. Are tank car linings visually inspected? How? Is this satisfactory?

d. Are small package containers, or fiber or steel drums if used, internally inspected for cleanliness prior to filling?

e. How do corrugated cartons and fiber drums hold up in storage? Do they sag or crease? Are they clean, or shop-worn, tired looking or dirty in storage?

f. Are new containers (drums, bags, bottles, etc.) being used in production on a FIFO (first in, first out) basis?

I. Off-Specification and Off-Weight Material

a. How is such material handled?

b. Is off-specification material reworked? Dumped? Shipped anyway? Who decides?

c. Is off-weight material brought up to weight?

d. What precautions are being taken in the plant to prevent off-specification or off-weight product from being shipped? Is precaution adequate?

e. What is procedure for checking accuracy of scales or small package weighing equipment?

J. Packaging and Labels

a. Are containers and packages covered by appropriate specifications?

b. Are containers approved for intended use?

c. Have labels been approved for use? Do they meet all legal requirements?

d. Do they show establishment number? EPA registration number?

K. Quality Assurance - General

   a. Is food grade or USP material segregated from standard
       production in warehouse?

   b. Is toxic material segregated from non-toxic material in
       warehouse?

   c. Is warehousing of finished goods orderly, clean and dry?

   d. Have you asked the question of plant people: Is it always
       like this? or, Is it always done this way? (The answers
       will frequently amaze you.)

L. Environmental Auditing (where applicable)

   a. Who took the sample? When, where, how?

   b. Does this conform with EPA, state, county, or municipal
       requirements? Is anyone responsible for seeing to this?

   c. Who took the flow rate data? How, when, where? Is this
       data reliable?

   d. Does the time of taking flow data coincide with sampling
       time? If not, has EPA approved variance in permit?

   e. Are laboratory data entered in ink in hardbound, sequen-
       tially-numbered-page notebooks used exclusively for the
       purpose?

   f. Who did the calculations for reporting purposes?

   g. Are calculations shown in notebook?

   h. Have calculations been checked by several persons before
       entry in report?

   i. Are report data (NPDES) exactly the same as laboratory
       data?

   j. Have all samples taken been reported?

   Admittedly, this represents a considerable amount of ground
to cover. On the other hand, it need be done only once or twice
a year in its entirety.

QUALITY-ASSURANCE AUDIT CHECKLIST

Audit By _____

Location _____   Date _____

	O.K.	Other	Notes
1 Analytical methods written up			
2 Analytical methods available in lab			
3 Data recorded in notebooks			
4 Data recorded in ink			
5 Lab data retention (6 yr. min)			
6 Copy of current Mfg. Spec. in lab -office			
7 Quality control reporting forms available and used			
8 Condition of lab			
9 Condition of lab instrumentation			
10 Condition of lab glassware			
11 Lab balances, physical condition			
12 Balances checked and calibrated			
13 Retained-sample area			
14 Retained-sample period			
15 Track retained sample to notebook			
16 Raw materials analyzed			
17 Off-spec product notification			
18 Off-spec product disposition			
19 Track retained sample through shipping			
20 Raw materials purchased per spec.			
21 Plant lab-technician training			
22 Condition of plant lab			
23 Instruments in plant lab			
24 Glassware in plant lab			
25 Plant-lab notebooks			
26 Plant-lab data in ink			
27 Product sampling by trained personnel			
28 Warehouse conditions			
29 Product label			
30 Product lot marking			
31 Container weight accurate			
32 Packaging scales condition			
33 Packaging scales checked and calibrated			
34 Condition of loading area			
35 Tank car/truck cleanliness			
36 Tank car/truck inspection			
37 Tank car/truck marking			

Figure 4. Suggested check list for use in conducting periodic audits of procedures used in quality control. (Reprinted by special permission from CHEMICAL ENGINEERING, February 11, 1980, by McGraw-Hill, Inc. , New York, N. Y. 10020.)

To simplify things a bit, refer to Figure 4. This is a Quality Assurance Audit Checklist. Armed with copies of this (or a list equally as suitable to your needs), quality control personnel can conduct a reasonably quick (and adequate) audit — and do it at relatively frequent intervals.

A similar type of checklist can be designed for warehouse quality audits. Such a checklist might include the following items:

1.   Action taken in event of emergency, spills, breakage, etc. Does warehouse have adequate information to handle?

2.   Retain packing lists of car/truck shipments from other branches of your company or suppliers.

3.   Store unsaleable items separate from saleable goods.

4.   Assign consecutive warehouse lot number to each unbound shipment and note this number on all copies of warehouse receipt.

5.   Are open-topped trucks or trailers used for shipments? Are tarpaulins available?

6.   Are outbound shipments secured, braced or dunnaged properly?

7.   Are oxidizer labels available for shipments requiring them? Trucks need them for shipments over 1,000 lbs.

8.   Are containers or drums patterned properly on pallets? Are they stacked in accordance with sidewall strength of container?

9.   Are damaged, defective, or deteriorated packages stored separately and safely?

10.   Is there a procedure for repackaging or disposing of such packages?

11.   Is product storage cool, dry and well ventilated?

12.   Are products stored away from incompatible materials?

13.   Is stock rotated on first in - first out basis?

14.   Does warehouse have written procedure for disposal of small quantities of spilled product? Large quantities?

15.   Are overpack drums, or repack bags (or similar containers) kept on premises. Is it legal to repack product at this warehouse?

Of course, you may add to or delete from this list. However use a list — don't wander aimlessly through a warehouse, just looking. This is sightseeing, not an audit.

8
PRODUCT LIABILITY

I.   INTRODUCTION

In recent years, there has developed a sharply rising trend in the
incidence of product liability cases being brought to court. Recent
changes in the laws dealing with the legal responsibilities of the
manufacturer have led to rulings by the courts that are greatly
expanding the circumstances in which a consumer can collect
damages for losses suffered from either a breach of warranty or
negligence on the part of the manufacturer.

Equally important are the growing indications that federal and
state governments will more frequently prosecute manufacturers
for producing defective products or using manufacturing processes
which could be injurious to life and health.

In general terms, "product liability" refers to the legal responsibility of a manufacturer to compensate a consumer who has
been adversely affected by the use of the product.

Product liability, under current court rulings, can be caused by
negligence or breach of warranty. Both can apply to the chemical
process industries.

The recent spate of suits against chemical manufacturers who have dumped toxic wastes in approved sites close by what were, or have since become, inhabited areas, have, as their main contention, the dumping of the chemical wastes, and alleged lack of attendance to the site in later years. This, despite the fact that no law was broken at the time of the use of the dump site.

This seemingly stretches the interpretation of negligence which heretofore described the manufacture of a product which has been proven to be defective because of poor design, materials or quality.

Breach of warranty describes the failure of a product, in this case, a chemical, to perform as represented by the producer. Such representation may include sales specifications, advertising and sales literature. So, not only is the manufactured product a source of risk, everything written about it contributes to the risk.

No company can afford to ignore the threat of product liability lawsuits. It is a subject of concern because of the attitude of legal agencies and courts and the size of judgments awarded on settlements made out of court.

It is also of concern because rapid introduction of a product into the market may cause a very real problem of multiple claims before recall can be initiated; and a recall can be time consuming and very costly.

The trend toward increasing product liability suits is irreversible. It is prudent, therefore, to examine your present position with an eye to minimizing your exposure.

There is no easy solution to the problem and complacency about the lack of negligence in your operations can be dangerous.

The chemical manufacturer, today, is faced with two major problems in this area. First, claimants do not necessarily have to prove negligence; in fact, the claimant's own negligence in the use of your product may have little bearing on the outcome of the case. Second, and it is worth repeating, liability suits may be based on the way the product is packaged, distributed, advertised, labeled, or described in sales specifications and literature.

To underline your vulnerability, your product need not even have been purchased or used by the claimant.

The employees of a chemical company who were engaged in cleaning a tank car which had contained ammonium hydroxide, pumped water into the car. The escaping fumes caused paint on the claimant's house to discolor and peel. He lived a half mile away. His suit was successful.

In another case, an infant brought into a store by his grandparents, sued for injuries from an exploding aerosol product on a shelf. The infant's suit was successful too.

It is not only the affluent large corporation which may be sued; a small company may also end up in court. Smallness will not discourage potential suits.

Improper packaging cost a small household chemical manufacturer a million dollars. A tightfitting screw cap on one of his containers permitted gases to collect and build to explosive pressure within the can. The can exploded in a housewife's face and blinded her.

II. COMPANY POLICY, PRECAUTIONARY MEASURES, AND QUALITY CONTROL

All is not lost, however. You must develop a company-wide plan of action for minimizing exposure to product liability suits.

There are a variety of precautionary measures which can and should be taken.

1. Develop a company policy which reflects management's commitment to make the finest product possible. This doesn't necessarily mean the most expensive. It is important for plant personnel to know that off-specification product will not knowingly be shipped just to maintain production quotas.

2. Develop an effective quality control program. Such a program should have as its first requisite, staffing with trained, competent personnel. In all too many plants, the laboratory staff is at the lower end of the salary and stature scale. They may have minimal clout and their decisions may frequently be overriden by production or marketing personnel.

    In many plants where the the laboratory is part of a bargaining unit, low scale, untrained plant personnel may opt for a laboratory opening if only to obtain an indoor job during winter.

    If you accept less than highest standards for hiring laboratory personnel, you will doom your quality control program from its very inception.

Insist on production quality control. It is not enough to
have manufacturing specifications. Each operating station
should have copies of specifications and other product information pertinent to its role in the manufacturing
scheme.

On-line process control instruments should be checked or
calibrated for accuracy at suitable intervals. Maintenance
of production equipment demands more than lubrication
and replacing an occasional nut, bolt or motor. Prevention of rust formation, or the replacement of rusted equipment, part or whole, is essential. Rust particles in your
product may cause all manner of problems, particularly
in consumer products.

3.   Develop a records retention program. Save all your pro-
     duction data, laboratory analytical data, and shipping
     documents. Don't be in a hurry to throw these things
     away merely because you are accumulating large quanti-
     ties of paper. The records you dispose of today may well
     come back to haunt you in several years.

4.   Double-check your labeling program. Products must be
     labeled in accord with state and federal laws; national
     and local codes and standards set by trade organizations.
     Maintain a label history file by product.

5.   Check all sales literature and advertising claims for accu-
     racy. Courts and federal agencies are now tending to
     rule that these are extensions of product warranties.
     Avoid the use of absolutes in your literature, such as
     "absolutely harmless" or "guaranteed." Be certain you
     are technically correct and can verify, with sound data,
     any statement you make. Wherever possible, use legal
     counsel to certify the legality of your wording.

6.   Don't ignore your packaging. Correct packaging is as
     important as correct labeling. Examine your containers,
     drums (steel, fibre or plastic) drum linings, caps,
     spouts, cartons, inserts, reshippers - all the packaging
     materials and components you use. A bottle cap which is
     too tight, or loose, can cause untold problems. If your
     product requires a child resistant closure, be certain it
     passes the approved protocol for such closures. The
     entire package should be designed with the type of pro-
     duct and the mode of shipment in mind.

7. Review your guarantees, warranties and disclaimers. If you are unfamiliar with pertinent law or recent court rulings, consult counsel with competency in this area. Limited warranties and disclaimers cannot necessarily be relied upon to relieve you of liability claims.

8. Develop a standby product recall system. In the unhappy event you are required to recall a shipment, know where the product went and how it can be identified. As we have discussed in another chapter, identification is made easier by having a date code and lot number on each individual container and noting this date code and lot number on the relevant documents for this shipment.

9. Review your complaint and claims handling procedures. It may be possible to minimize or, in some cases, even prevent damaging and expensive litigation. It is not only the financial settlement which is involved but the publicity which can be even more damaging. You will recall the painful experience of Bon Vivant soups, which was forced out of business by bad publicity caused by a recall.

   a. Speed is of the essence in handling a claim. Get all the information while it is still fresh.

   b. Consult with a competent attorney.

   c. Notify your insurance carrier promptly.

   d. If possible, obtain a sample of the product in question. Better still, try to obtain the complete package. Don't rely upon it being shipped back to you by the potential claimant. Send somebody to get it, quickly. Determine that the product is, indeed, yours. Check the product code on the container, sales records, shipping documents. Don't leave anything to chance and take nothing for granted. Make no assumptions. If acceptable to the claimant, try to have him accept replacement product. This could soothe him and minimize your problem.

   e. Have your quality control laboratory analyze the returned product and compare it against your retained production sample and the product specification. If you use the facilities of an outside laboratory, consult with your counsel before sending the sample to them for analysis. It is possible that the results of the

outside laboratory could be made available to a potential claimant during the discovery procedure. Should
the data be adverse to your company, this could be
most damaging to your defense. Consult with your
counsel on this.

f. Make no admissions to the claimant or the attorney
    representing him. Your counsel should represent
    you.

g. Don't offer a settlement just to get the claimant off
    your back. This could be tantamount to an admission
    of culpability where none may exist.

I mentioned the "discovery procedure" above (e). It is the
procedure in litigation which gives to your adversary access to
your correspondence, data, files, records; indeed, anything and
everything he may deem pertinent to his case.

III. RECORD KEEPING

Despite the fact that your records may be available to a claimant
through discovery, a good recordkeeping program may well be
your best defense. Well kept production records and quality control data on raw materials, containers and finished product are a
sign of your integrity and desire to provide your customers with
a satisfactory product.

As indicated in a previous section of this book, all records,
at a minimum, should be retained for the useful life of the product.
This should include whatever time period may be considered suitable for all product of a particular lot to be used completely by
any potential customer.

In light of the present legal and regulatory climate, a twenty
year record retention period is something which should be accorded
serious consideration. Storage space for these records should not
be a factor in your consideration.

IV. PRODUCT RECALLS

Allied with the problem of product liability suits is the problem
of product recall. It is safe to say that one of the most serious

situations that can beset a company concerns the recall of a product because of a safety or quality problem.

Should you be ordered by a federal agency to issue a recall; or should your company make such an emergency decision, it will become quickly obvious that, in this unhappy circumstance, accurate laboratory data and adequate production and/or shipment samples, satisfactorily retained, form the first line of defense.

There is, at least, one other potential source of information that can help make a recall situation less of a nightmare. This is the shipping document that is usually retained by the plant's shipping and receiving department. It may be a copy of the purchase order or a shipping instruction. It is, in any event, a document which authorizes shipment of product to the customer. It states, among other things, how much of what is to be shipped to the customer.

Now comes the important part which bears upon your recall situation.

On this sheet, this shipping document, should also be noted, by the shipping crew, the batch or lot number of each item shipped as it is loaded into the shipping vehicle. This means every drum, container, carton, whatever. No exceptions.

Even if many different batch or lot numbers are represented in the shipment, it is necessary that this information be shown for each package in the shipment.

If your product is being shipped to a warehouse or distribution point, it is still valid to include such information. It follows that it is important for batch or lot numbers to be included on any shipping document used to release product from the warehouse or distribution point to the customer.

Such documents are then extremely helpful in tracing specific lots of material subject to recall. Imagine trying to find such product without any knowledge of where it was shipped.

Since product recalls may increasingly become an unhappy fact of business life, it will become a worthwhile practice to schedule a "dry run" recall procedure once in a while, until a reasonably satisfactory program has been developed.

For example:

Assume a production batch conatained 50,000 lbs. of product which was then packaged in 50 lb. Kraft bags. This total of 1,000 bags was then shipped out to customers, distributors and warehouses. If shipping papers had been filled out as suggested above, a search of these papers, however tedious, would reveal where these bags were and how many were still available for recall.

An interesting and useful audit to run every few months involves selecting a batch number of any product, at random, from a shipping document and tracing it backwards: Find the retained production or shipment sample for that code and then go back to the retained note book to find the data pertinent to the sample.

If this cannot be done smoothly each time, there is a weakness in your system. Find it and improve it.

SUGGESTED READING

Burr, Irving, "Elementary Statistical Quality Control," 1978, Dekker.

Burr, Irving, "Statistical Quality Control Methods," (Statistics: Textbooks and Monographs, Vol. 16) 1976, Dekker.

Case, Kenneth E. & Jones, Lynn L., "Profit Through Quality: Quality Assurance Programs for Manufacturers," 1978, American Institute of Industrial Engineering.

Charbonneau, Harvey C. & Webster, Gordon L., "Industrial Quality Control," 1978, Prentice-Hall.

Crosby, Philip B., "Quality is Free: The Art of Making Quality Certain," 1979, McGraw-Hill.

Grant, Eugene L. & Leavenworth, Richard, "Statistical Quality Control," 5th ed., 1979, (With Solution Manual), McGraw-Hill.

Juran, Joseph M., "Quality Control Handbook," rev. 3rd ed., 1974, McGraw-Hill.

Juran, Joseph M. & Gryna, Frank M., Jr., "Quality Planning & Analysis From Product Development Through Use," 2nd ed., 1980 (with Solutions Manual) McGraw-Hill.

Kateman, G. & Pijpers, F. W., "Quality Control in Analytical Chemistry," 1981, Wiley.

Matley, Jay & Chemical Engineering Magazine, "Practical Process Instrumentation and Control," (Chemical Engineering Series) 1980, McGraw-Hill.

Shinskey, F. G., "PH & P Ion: Control in Process and Waste Streams," (Environmental Science & Technology Series) 1972, Wiley.

Squires, F. H., "Successful Quality Management," 1980, Hitchcock Publishing Co.

Journals and Papers

American Society for Quality Control, *Annual Technical Conference Transactions*, Thirtieth, 1976, ASQC, Milwaukee, Wis., June 1976, 588 pp.

American Society for Quality Control, *Annual Technical Conference Transactions*, Thirty-first, 1977, ASQC, Milwaukee, Wis., June 1977, 618 pp.

American Society for Quality Control, *Annual Technical Conference Transactions*, Thirty-second, 1978, ASQC, Milwaukee, Wis., May 1978, 714 pp.

Anon., "When Quality Control is a Part-time Job," *Business Week*, n 2613 (Industrial Edition) Nov. 1979.

Bader, M. E., "Quality Assurance and Quality Control," *Chemical Engineering*, Vol. 87, Feb. 1980, pp. 86-92; Apr. 1980, pp 89-93; June 1980, pp 123-129; Aug. 1980, pp 95-97.

Fereday, R., "Quality Control — A Management 'Must Have',"" *Productivity & Technology* (New Zealand) May 1980, pp 4-7.

Geoffrion, L. P., "Contribution of AC to Product Liability Aversion," *Transactions of Annual Technical Conference*, ASQC, Thirty-second, May 1978, ASQC, Milwaukee, Wis., 1978.

Lakshmikanthan, P. R., "Quality Assurance in Liquid Filling Operations," *Quality and Reliability Journal*, Vol. 4, No. 1, Jan. 1977, pp 17-18.

Lawlor, A. J., "Quality Assurance in Design and Development,"
   Quality Assurance, Vol. 4, No. 3, Sept. 1978, pp 87-91.

Natelson, D. M., "Quality Assurance, A Primary Management Tool
   for Product Liability Prevention," *Proc. Prod. Liability Prev.
   Conf.* PLP-77E, Aug. 22-24 1977, pp 53-63.

Schock, H. E., Jr., "Regulatory Cost Management and Product
   Assurance," *Proc. Annu. Reliab. Maintainability Symp.*,
   Washington, D.C., Jan. 1979, Publ. by IEEE, New York, NY,
   1979, pp 13-17.

Law Journals

Calabresi, "Products Liability: Curse or Bulwark of Free Enter-
   prise," *27 Clev. St. L. Rev.*, 313, 1978.

Croyle, "Impact Analysis of Judge-Made Products Liability
   Policies," *13 Law & Soc. Rev.*, 949, 1979.

"Evolving Consumer Safeguards — Increased Producer and Seller
   Responsibility in the Absence of Strict Liability," *5 U. Balt.
   Law Rev.*, 128, 1975.

Fischer, "Products Liability — Functionally Imposed Strict Liability," *32 Okla. Law Rev.*, 93, 1979.

Frank & Ringkamp, "Products Liability Primer," *23 Prac. Law*, 75,
   1977.

Ghiardi, "Products Liability — Where is the Borderline Now?,"
   13 Forum, 206, 1977.

Henderson, "Extending the Boundaries of Strict Products Liability:
   Implications of the Theory of Second Best," *128 U. Pa. Law
   Revl*, 1036, 1980.

Kroll, "Products Liability - A Reasonable Expectation - The Ultimate Goal," *25 Drake Law Rev.*, 828, 1976.

Phillips, "A Synopsis of the Developing Law of Products Liability,"
   28 Drake Law Rev., 317, 1979.

"Products Liability: Tort or Contract — a Resolution of the Con-
   flict," *21 N.Y.L.F.*, 587, 1976.

Twerski, "From Defect to Cause to Comparative Fault — Rethinking
   Somer Product Liability Concepts," *60 Marq. Law Rev.*, 297,
   1977.

Wade, "A Conspectus of Manufacturers' Liability for Products,"
     10 Ind. Law Rev., 755, 1977.

Walkowiak, "Product Liability Litigation and the Concept of Defec-
     tive Goods: 'Reasonableness' Revisited?," *44 J. Air Law*, 705,
     1979.

Wallace, "Products Liability — Current Developments and Direc-
     tions," *43 Ins. Counsel J.*, 519, 1976.

_____, "Product Liability Disease Litigation: Blueprint for
     Occupational Safety and Health," *16 Trial*, 25, 1980.

INDEX

141